Mycobacteria

Institute of Medical Laboratory Sciences Monographs

Under the General Editorship of

F. J. BAKER, OBE, FIMLS
Formerly Principal Medical Laboratory Scientific Officer,
Brompton Hospital, London

Titles currently available:

Medical Laboratory Statistics by Paul W. Strike
Laboratory Control of Antibacterial Chemotherapy
  by Michael Bryant
Mycobacteria by Maureen V. Chadwick
Lymphoid Tissue by G. J. Reynolds

Some forthcoming titles:

Laboratory Investigation of Urine by Ian Leighton
Computer Systems in Medical Laboratory Sciences
  by Paul Ayliffe and Martin Walter
Microbiology in Blood Transfusion by John A. J. Barbara

Mycobacteria

Maureen V. Chadwick FIMLS
*Chief Medical Laboratory Scientific
Officer, Department of Pathology,
Brompton Hospital, London*

WRIGHT · P S G

Bristol London Boston
1982

Published by:
John Wright & Sons Ltd, 823–825 Bath Road, Bristol BS4 5NU,
England
John Wright PSG Inc., 545 Great Road, Littleton, Massachusetts 01460,
U.S.A.

First published, 1982
Reprinted, 1982

British Library Cataloguing in Publication Data

Chadwick, Maureen V.
  Mycobacteria.—(Monographs in medical
  laboratory science series)
  1. Mycobacterium
  I. Title   II. Series
  589.9′2   QR82.M8

ISBN 0 7236 0595 5

Library of Congress Catalog Card Number:
81–52685

Printed in Great Britain by
John Wright & Sons Ltd, at the Stonebridge Press, Bristol BS4 5NU

Foreword

With the rapid advancement of medical laboratory sciences, it is often difficult for the laboratory worker to keep abreast of current knowledge. Major textbooks cover a wide field of a given major discipline, but obviously they cannot cover every facet of a subject.

The Institute of Medical Laboratory Sciences Monograph Series is written by experts in specific areas, to expand on a subject and to give it the depth and breadth not generally possible in a major work. These books will be useful to those working in the medical laboratory sciences field, for use on the bench, as an examination reference book and for keeping up with current thinking.

I am delighted to have been offered the General Editorship of this series and hope it will be the success it well deserves to be. It is hoped that the Monograph series will be extant for a long period of time and that it will build up to be a useful and welcome addition to your library.

F. J. Baker
London 1982

Preface

Although tuberculosis is no longer the major disease it was at one time in the United Kingdom, it remains a serious problem requiring specialised isolation, identification and treatment. In the Third World, tuberculosis is still a disease of major proportions and it is for this reason, and for the now identified large numbers of disease producing mycobacteria, that this book was written.

Not every laboratory will cover all aspects of mycobacterial identification, but help is given here for those who need to progress beyond the microscopy of acid-fast bacilli. It is intended that this book will be of use to examination candidates as well as to those working at the bench. I wish to acknowledge with thanks, the help and encouragement received from Frank Baker O.B.E., not only during the preparation of this book but throughout my career.

Maureen V. Chadwick
London 1982

Contents

1 Tuberculosis–the Killer Disease

Tuberculosis kills one person every 30 seconds somewhere in the world today. 'Captain of All the Men of Death' is an apt name for this killer disease which has been a dark shadow over mankind for centuries—indeed evidence of its existence has been found in Egyptian mummies.

Tuberculosis is commonly a disease of the lungs (pulmonary tuberculosis), although it can take many extrapulmonary forms, such as lymphadenitis, bone and joint disease, subcutaneous cold abscesses and meningeal tuberculosis. Cervical lymphadenitis is a disease which was once common in young children, with the tonsils as the primary focus of infection. Today this is seen commonly in Asian immigrants as a secondary infection rather than a primary complex. Genitourinary tuberculosis is a disease which is not always diagnosed early enough, as microscopy on the urine samples is not always positive. In most cases tuberculous infection of the female genital tract is secondary to the primary focus elsewhere in the body. Dissemination may take place by the blood stream from an extra-abdominal focus situated in the lungs, the lymphatic glands, the bones, joints or urinary tract, or by direct local extension from a focus within the abdominal cavity.

Although it is no longer the most common cause of death in countries with a high standard of living, it is still the number one killer throughout the world. Worldwide, it is estimated that there are 10 000 000 people suffering from the disease and in the United Kingdom alone there are over 10 000 cases annually, despite the advances that have been made in chemotherapy.

HISTORICAL REVIEW

The Search for a Cause

During the nineteenth century, some years before Robert Koch's discovery of the tubercle bacillus, a famous British surgeon, Sir William Budd, recorded that when talking to Dr Livingstone he had been told that tuberculosis was almost unknown in the African interior. However, the disease was appearing in the central trading areas, and a South Sea Island's population had been entirely wiped out by tuberculosis. Moreover, Budd had treated among his patients at Bristol, Negroes who had contracted the disease whilst at sea, working on British ships. In 1856 Budd postulated the idea that man infected man and that it appeared that Europeans were responsible for infecting Polynesians, Indians, Africans and Eskimos when discovering and trading with these people.

One of Budd's contemporaries, a French army surgeon, Jean Antoine Villemin, took a fundamental step forward by demonstrating that tuberculosis could be transmitted from man or cow to the rabbit and guinea-pig and that transmission from one infected animal to another could go on in unending series. His experience as a military surgeon made him aware that soldiers stationed for long periods of time in army barracks were more susceptible to tuberculosis than troops in the field, and that healthy young men from the country districts often became consumptive within one to two years of their arrival in army barracks.

Villemin showed that crowded living conditions were a major contributing factor to the contracting of tuberculosis. His experiments demonstrated beyond doubt that tuberculosis did not originate spontaneously but was caused by a germ multiplying in the body and transmissible to a well person by either direct contact or by infected droplets in the air. He also reported that the disease was enhanced as a result of emaciation, physiological misery, atmospheric disturbances, bad heredity, unhealthy occupations or prolonged debilitating maladies. He presented the discovery for the first time to the French Academy of Medicine in 1865, but his reports were received with such indifference that not even William Budd became aware of them. The medical profession at that time believed in an innate susceptibility to tuberculosis and were not receptive to the theory of contagion. Villemin published his work in 1868, and slowly his germ theory of disease gained widespread acceptance.

In 1882 three German workers saw the tubercle bacillus in infected tissues independently and almost simultaneously, but the work performed by Robert Koch was so expertly presented to the Physiological Society of Berlin on 24 March 1882 that today the entire glory is given to him. He was the man who convinced the world that tuberculosis was a contractable disease, and thus the work of Villemin was ignored. Robert Koch became a national hero in Germany and was given new laboratories so that he could work on the tubercle bacillus. He was acclaimed throughout the world, creating the feeling that man had finally come to grips with the greatest killer of the human race. The 'Captain of All the Men of Death' was no longer a vague phantom.

The Search for a Cure

In 1890 Koch announced to the Tenth International Congress of Medicine in Berlin that he had discovered a substance which could not only protect against tuberculosis but could even cure the established disease. It was merely a glycerin extract of tubercle bacilli, today known as 'Old Tuberculin'. This news gave people throughout the world hope, and many rushed to Germany to be treated by Koch. It soon became obvious that tuberculin killed more patients than it helped, and the treatment fell in to disrepute almost everywhere. One year after the glowing reports of Koch's discovery of tuberculin, the *British Medical Journal* published a scathing article condemning Koch for his unethical behaviour. Nevertheless, Koch's work had not been in vain, for tuberculin was later to prove a useful tool in the detection of tuberculosis.

Throughout the nineteenth century, many famous people died from tuberculosis or consumption as it was then called.

The tragedy of consumption led to a perverted romantic attitude towards the disease. Keats and Shelley the poets both died from tuberculosis and wrote poetry during their slow deaths. Lord Byron had a morbid desire to die from consumption because the ladies would then say, 'Look at poor Byron, how interesting he looks in dying'.

In the United Kingdom, during Queen Victoria's reign, 1 in every 5 people died from tuberculosis compared to 1 death in 1000 today. Better social conditions have played a major role in the containment of the disease.

During this period many cures were suggested for the treatment of tuberculosis, such as the inhalation of vapours from resinous balsamic substances and herbs and the use of opiates for quietening coughs. Most had little or no effect on the course of the disease but owed their lasting popularity to the fact that they improved the sense of wellbeing of the patients. Opium was also widely used in the nineteenth century, again giving relief but not a cure.

Cod-liver oil was a new treatment introduced by Dr C. J. Blasius Williams, one of the founders of the Brompton Hospital for Consumption. In 1850 Williams stated that 'pure fresh oil from the livers of the cod is more beneficial in the treatment of Pulmonary Consumption than any agent, dietetic or regimenial, that has yet been employed'. This treatment certainly increased the weight of the patient by providing additional vitamins A and D to their diet, but did little or nothing to help in curing the disease.

Much emphasis was placed on the benefits of the sanatorium, with the inclusion of rest, fresh air and good food. Many people suffering from consumption in the nineteenth century left Britain's damp cold climate to travel to the warmth of Italy in the hope of a cure. It is now known that these factors did very little to modify the outcome of the disease once contracted. However, diets and physique and health do play a part in the avoidance of the disease. For example, a survey in 1931 showed that in two African tribes, the Masai and Akikuyu, there was a marked difference in the incidence of tuberculosis. The Masai tribe were meat eaters and included milk and raw blood in their diet, whilst the Akikuyu were vegetarians. The incidence of pulmonary tuberculosis was far more prevalent in the vegetarians than in the meat eaters.

In addition to the medical cures being tried, surgical procedures for the treatment of tuberculosis were also carried out, the collapsing of the infected lung being tried by many surgeons. It was in 1882, the same year as the discovery of the tubercle bacillus, that Carlo Forlanini suggested that pulmonary tuberculosis could be treated by establishing artificial pneumothorax through the chest wall. This technique eventually became accepted by the medical profession. After realizing that more people could continue a reasonably active life with one lung, surgeons then started to treat certain forms of pulmonary disease by the removal of the affected lobe.

Chemotherapy

As social conditions and surgical techniques improved, more and more people escaped the terrible lingering death of tuberculosis, but men continued striving to find cures for all cases by use of the various chemotherapeutic agents. In the 1930s sulphanilamide was tried on tuberculous guinea-pigs with some success, although this was of no value in the treatment of humans. In 1943 streptomycin and *para*-aminosalicylic acid (PAS) were discovered and these showed immense activity against tubercle both *in vitro* and *in vivo*. Thus began the successful treatment of tuberculosis using chemotherapeutic agents. However, after the initial excitement caused by these two drugs, it was soon found that not all cures were permanent. Some relapsed as a result of the tubercle bacilli developing resistance to the drugs.

The search continued to find more drugs active against tuberculosis. In February 1952 a new miracle drug was announced—isonicotinic acid hydrazide (INH). This drug, used in combination with others, has proved to be the most effective antituberculosis drug to date. Of many drugs discovered subsequently, probably the most remarkable is rifampicin, which has enabled the length of treatment of tuberculosis to be reduced from the original 18 months to only 6 months and in some cases even less.

LOOKING TO THE FUTURE

Today, with a high standard of living and a battery of anti-tuberculosis drugs, most cases of tuberculosis are curable, yet in the United Kingdom in 1971, 52 of 265 people who died of active tuberculosis were not diagnosed until after death, and in most of the other deaths unsatisfactory medical care was considered to be a contributory factor. Although tuberculosis in the United Kingdom has reached a lower level today than in Queen Victoria's reign, continued vigilance against the disease is vital, in diagnosis, safety and prevention. In certain areas of the United Kingdom there has been a marked increase in the incidence of tuberculosis, and the level amongst laboratory workers is still a cause for concern. With the relaxing of mass radiography, BCG and routine screening, the question now to be considered is, have we become complacent?

2   Mycobacteriaceae

The most distinctive property of the Mycobacteriaceae is their characteristic staining. They are not readily stained, but once stained will resist decolorization by acids and alcohol. For this reason they are often referred to as acid-fast bacilli. They are strict aerobes, although growth is enhanced by an increased CO_2 tension, and grow slowly on special media at $37\,^{\circ}C$. The family contains a wide range of nutritional types, including saprophytic species that are present in the soil. The pathogenic members of the family cause some of the more important human infections, including leprosy and tuberculosis. Most species of birds, mammals and cold-blooded animals are susceptible to their own specific pathogenic species, although there is considerable cross-infection between animals and man.

The Mycobacteriaceae family contains a single genus, *Mycobacterium.* They are non-motile, aerobic, slightly curved or straight rods, $0\cdot2-0\cdot6$ by $1\cdot0-10\,\mu m$. Mycobacteria have a unique cell wall made up of complex lipids.

MYCOBACTERIUM TUBERCULOSIS

Tuberculosis is an ancient disease, having been demonstrated in the bones of some Egyptian mummies. It can infect many organs of the body.

M. tuberculosis (human tubercle bacillus) causes tuberculosis in man, and is sometimes transmitted to cattle, monkeys and dogs. Infection may be established in guinea-pigs but not in rabbits, voles or fowls. The bacillus is a slender, straight, or slightly curved rod with rounded ends. The organisms vary in width from $0\cdot2$ to $0\cdot5\,\mu m$ and in length from 1 to $4\,\mu m$.

7

They are acid-fast, non-motile, non-sporogenous and non-encapsulated. They are difficult to stain with the Gram stain, but are usually considered to be Gram-positive. They are best stained using the Ziehl–Neelsen stain or the auramine phenol stain. Organisms often have a beaded appearance because of their polyphosphate content and unstained vacuoles.

M. tuberculosis is a very slow growing organism even under optimal growth conditions. A medium usually contains serum or egg, mineral salts and a carbon source, e.g. glucose glycerol or pyruvic acid. *M. tuberculosis* will normally grow in 14–28 days; if, however, the patient has been subjected to chemotherapy then the organisms may take up to 8 weeks to grow.

M. tuberculosis will appear as buff-coloured, small dry colonies often referred to as being 'breadcrumb'-like. Under optimum culture conditions, the organism's doubling time is 18–24 hours. The addition to broth medium of a surfactant, such as Tween 80, improves the growth rate of the organisms and, by helping to prevent aggregation of cells, permits them to grow as a smooth suspension.

Tubercle bacilli are highly resistant to drying. Cultures maintained at 37°C have been found both viable and virulent after storage for 12 years. The environment in which bacilli are found is an important factor in their viability. Organisms from cultures will die in 2 hours when exposed to direct sunlight, but bacilli contained in sputum require an exposure of 20–30 hours before they are killed. When protected from direct sunlight tubercle bacilli can survive for several months in sputum.

MYCOBACTERIUM AFRICANUM

The properties of *M. africanum* were first described by Castets and others in 1969, and are intermediate between those of the human and bovine types. It is sensitive to pyrazinamide, produces more niacin than *M. bovis* and is less dependent on sodium pyruvate for growth. *M. africanum* has been referred to as the Afro-Asian variety of *M. bovis;* it is most commonly found in the immigrant population.

MYCOBACTERIUM BOVIS (bovine tubercle bacillus)

Mycobacterium bovis was differentiated from *M. tuberculosis* by Theobald Smith in 1896. This species is often shorter and plumper than the human tubercle bacillus and its primary isolation is more difficult. *M. bovis* is the causative organism of tuberculosis in cattle, which is transmissible to man, monkey, pig and domestic animals. It is more highly pathogenic for animals than *M. tuberculosis*. The organisms grow more slowly and on Lowenstein–Jensen medium give a dysgonic growth with small flat colonies. However, if sodium pyruvate is added to the media the growth of *M. bovis* is enhanced.

Fifty years ago most dairy herds were heavily infected with bovine tuberculosis and the milk from these animals provided a common source of disease in man. Tuberculin testing of cows and pasteurization of milk has greatly reduced the incidence of bovine tuberculosis in man.

Bacille de Calmette–Guérin (BCG)

This is a variant of *M. bovis* isolated in 1908 by the French workers, Calmette and Guérin, attenuated by repeated subculture on a glycerol-potato-bile medium. Subcultures of the original isolate are maintained as *M. bovis* strain BCG and used as an immunizing agent against tuberculosis. Although regarded as a very safe vaccine, it occasionally causes local abscesses or more widespread infections, including osteitis following vaccination in the newborn.

MYCOBACTERIUM MICROTI (Murine tubercle bacilli)

M. microti is responsible for tuberculosis in voles, commonly referred to as the vole bacillus. It grows more slowly than either the human or bovine organisms and tends to be longer ($10\,\mu$m) and thinner. Primary growth does not occur on media containing glycerol, nor does glycerol enhance growth on subculture.

M. microti is immunologically closely related to *M. tuber-*

culosis and *M. bovis*. For this reason it was included in pilot studies with BCG to test its effectiveness when compared with BCG in the immunization of humans against tuberculosis. A level of protection was provided after 7·5–10 years which was practically identical to that provided by BCG.

MYCOBACTERIUM ULCERANS

This organism produces a destructive, primarily tropical skin disease which produces chronic ulcers with necrotic centres if not treated early. The bacilli are 0·5 μm wide by 1·5–3·0 μm long, but in tissue sections they are usually larger and beaded. *M. ulcerans* is a very slow growing organism with an optimal temperature for growth of 33 °C; it may take up to 12 weeks to grow. The organism produces small domed colonies which are smooth and pale cream to yellow in colour.

Most laboratory animals are resistant to infection by *M. ulcerans*, although the footpads of mice may be infected by injection of the organisms.

M. ulcerans infection has been most frequently encountered in isolated pockets in Australia, Africa and Mexico, with close proximity to rivers and swampy areas. The organism has never been demonstrated outside the human body.

MYCOBACTERIUM LEPRAE (Hansen's bacillus)

Hansen, in 1874, described the presence of myriads of bacilli in the lesions of leprosy patients. Although it is over 100 years ago that this organism was first mentioned it still remains a mystery. All attempts to culture *M. leprae in vitro* have failed, and it is difficult to infect experimental animals. A large amount of experimental work has been done on the footpads of mice. The nine-banded armadillo is the animal found to be most susceptible to *M. leprae* and studies are at present being carried out using this animal. In this animal the disease becomes disseminated to all organs, in contrast to the disease in man. The armadillo is useful for immunological studies and for producing large numbers of *M. leprae* for laboratory investigations. In infected human tissue it occurs as straight or slightly curved bacilli $2.8 \mu m$ in length and $0.2-1.4 \mu m$ wide, packed together in clumps in an arrangement that suggests a packet of cigars.

M. leprae is an obligate intracellular parasite which multiplies very slowly, man being the only natural host. The organisms are strongly acid fast, not alcohol fast, and sometimes show granular staining.

The clinical manifestation of leprosy varies enormously but forms a spectrum according to the immunological reactivity of the patient. At one pole of the spectrum (tuberculoid leprosy) there is a brisk response but at the other pole (lepromatous leprosy) there is no specific immune response to the pathogen.

In tuberculoid leprosy acid-fast bacilli cannot usually be demonstrated. Skin biopsies show mature granuloma formation in the dermis, consisting of epitheloid cells, giant cells and rather extensive infiltration of lymphocytes. The organisms invade the nerves and colonize the Schwann cells. The damage to the nerves is very extensive.

In lepromatous leprosy the histological picture is completely different to that of tuberculoid leprosy. Epithelioid cells and giant cells are absent, and lymphocytes are rare and diffusely distributed. The inflammatory infiltrate consists largely of histiocytes, with a unique foamy appearance due to an accumulation of bacterial lipids. Within the macrophages large numbers of acid-fast bacilli are seen. The nerves are also infected, with numerous bacilli in the Schwann cells. Damage to the nerve structure, however, is less than in tuberculoid leprosy.

MYCOBACTERIUM LEPRAEMURIUM (Stefansky's bacillus)

This organism causes rat leprosy, a disease that occurs spontaneously among house rats in many parts of the world. The disease is chronic and rats often live for 6—12 months after becoming infected. The organism in tissues of infected rodents simulates that of *M. leprae,* except that intracellular bacilli tend to be arranged at random rather than in bundles, and around rather than displacing the cell nuclei. Despite the similarities rat and human leprosy are totally unrelated. Rats are not a potential source of human leprosy.

The natural host for both *M. leprae* and *M. lepraemurium* is the monocyte. Unlike *M. leprae* however, *M. lepraemurium* can be maintained for months in tissue cultures of monocytes. Experimental infections can be established in rats, mice and hamsters.

MYCOBACTERIUM PARATUBERCULOSIS
(Johne's bacillus)

M. paratuberculosis is the cause of chronic specific enteritis of cattle occurring in Europe and North America. The disease is contracted by the ingestion of infected fodder and leads to diarrhoea and prevents the proper absorption of food. The incubation period is long and the progress of the disease is slow and invariably fatal. Skin testing of infected animals with bovine tuberculin is negative but is often positive with avian tuberculin.

M. paratuberculosis is a small, acid-fast rod, $1-2\,\mu$m long and $0.5\,\mu$m wide. It occurs in dense clumps, is strongly acid-fast and is either extra- or intracellular in distribution within the intestinal mucosa and submucosa. For many years growth could be obtained only by supplementing the medium with killed tubercle bacilli or the saprophytic *M. phlei.* The growth factor required by *M. paratuberculosis* is mycobactin, which it is unable to synthesize.

Primary cultures grow slowly, but subsequent subcultures, even on unenriched media, are more profuse.

OTHER MYCOBACTERIA

Many names have been given to these organisms in the past, including anonymous mycobacteria, which they no longer are, as all have been given names, non-tuberculous mycobacteria and opportunistic mycobacteria. The designation 'atypical' is an incorrect term as they are not atypical but are characteristic of their own particular species.

In 1955 Runyon proposed a very simple and useful scheme for classifying this heterogeneous group of mycobacteria based on their rate of growth and pigment production. This consisted of four groups, and within each group a number of organisms have been speciated. Collins (1962) and Marks and Richards

Table 1. Forty-one approved Mycobacterial names (Skerman et al., 1980)

M. africanum (HP)	M. gordonae	M. phlei
M. asiaticum	M. haemophilum (O)	M. scrofulaceum (O)
M. aurum	M. intracellulare (O)	M. senegalense (AP)
M. avium (O)	M. kamossense	M. simiae (O)
M. bovis (HP)	M. kansasii (O)	M. smegmatis
M. chelonei (O)	M. leprae (HP)	M. szulgai (O)
M. chitae	M. lepraemurium (AP)	M. terrae
M. duvallii	M. malmoense (O)	M. thermoresistibile
M. farcinogenes (AP)	M. marinum (O)	M. triviale
M. flavescens	M. microti (AP)	M. tuberculosis (HP)
M. fortuitum (O)	M. nonchromogenicum	M. ulcerans (O)
M. gadium	M. neoaurum	M. vaccae
M. gastri	M. parafortuitum	M. xenopi (O)
M. gilvum	M. paratuberculosis (AP)	

HP—human pathogen; O—opportunist; AP—animal pathogen.

(1962) have also classified these mycobacteria. Collins suggested 10 groups based on temperature requirements, growth on special media, pigmentation and sensitivity or resistance to thiosemicarbazone, whereas Marks and Richards suggested seven groups, based on Runyon's criteria, with the addition of temperature related to growth and sensitivity to several chemotherapeutic agents.

The atypical mycobacteria can cause tuberculous infections in man or animals, they are ubiquitous and have been found in practically every part of the world except Alaska. They appear to be endemic in certain geographical areas. The organisms have no known primary animal host but are environmental, occurring particularly in the soil. There is no evidence to suggest that these organisms can be transmitted from man to man. These mycobacteria generally invade a preformed lesion of the lung, e.g. pulmonary tuberculosis, carcinoma or an aspergilloma giving rise to an infection indistinguishable from tuberculosis.

The International Committee on Systemic Bacteriology (ICSB) has listed 41 named species of mycobacteria which have replaced or been amalgamated with any previous names given to various species. (Skerman et al. (1980), *Table* 1).

Runyon Group I. The Photochromogens

Photochromogens are those mycobacteria which become pigmented only when exposed to visible light. They are commonly associated with tuberculosis, simulating the growth of *M. tuberculosis* when grown at 37 °C in the dark, although they have smoother colonies. After 6–24 hours' exposure to light, they develop a bright yellow pigment.

M. kansasii (Hauduroy, 1955)

This organism is longer and wider than tubercle bacilli and often shows a banded appearance of stained and unstained parts. The organisms are usually arranged in curving strands and are acid and alcohol fast.

M. kansasii will grow on glycerol-containing medium in 2 weeks at 37 °C. If grown in the dark they will appear as off-white colonies very similar to *M. tuberculosis* but slightly smoother in appearance. On exposure to light they will develop a yellow colour.

Pathogenicity for animals is extremely limited, as in most laboratory animals only local lesions are produced.

M. kansasii is one of the most common atypical mycobacteria to infect man. It probably has the greatest pathogenicity for man than any other of the atypical mycobacteria.

M. kansasii produces pulmonary and extrapulmonary disease that is almost indistinguishable from that produced by *M. tuberculosis.*

M. kansasii will not grow on sodium pyruvate medium in some cases.

M. marinum (Aronson, 1926)

There have been three organisms described which have similar characteristics, *M. balnei, M. marinum* and *M. platypoecilus*, but they are now incorporated under the one name, *M. marinum* (Kubica, 1978). This organism is the cause of swimming pool granuloma. Its optimum temperature for growth is 31 °C, which enables it to be differentiated from *M. kansasii*. In tissues it may be seen in clumps as short, thick and uniformly staining rods, or long, thin, beaded and barred bacilli scattered throughout the tissue.

Culture at 31 °C on Lowenstein–Jensen medium produces soft, greyish white colonies. Upon exposure to light at room temperature the colonies develop an intense yellow orange pigmentation.

Although for years *M. marinum* had been isolated from tuberculous lesions of salt-water fish, its association with human infection was first described by Linell and Norden (1954). These workers isolated the organism from skin lesions of patients during an epidemic in Sweden. The source of infection was established as bacterial colonies growing on the rough cement of swimming pools and tropical fish tanks.

M. simiae

This organism was first described by Karassova et al. in 1965. Certain biochemical and cultural properties, including its ability to produce niacin, distinguish it from other photochromogens. It produces small dysgonic colonies which, when exposed to light, pigment yellow very slowly. Unless pigment tests and the niacin test are performed it could be confused with *M. intracellulare*.

Disease in man has not been well defined. It is, however, quite probable that infected rhesus monkeys, from which the organism was first isolated, might transmit the mycobacterium to man.

M. simiae is resistant to most of the usual antituberculosis drugs. Pyrazinamide, cycloserine, capreomycin and kanamycin should be tested, as well as tetracycline and erythromycin.

Runyon Group II. The Scotochromogens

The scotochromogenic organisms produce yellow or yellow-orange pigment in the dark, and show a deeper orange or reddish pigment if grown in continuous light. The colonies are smooth and generally grow between 2 and 3 weeks at 37°C.

M. scrofulaceum

This organism was identified by Prissick and Masson (1956) as the aetiologic agent of a form of lymphadenitis in children.

M. scrofulaceum cells are longer, thicker and more coarsely beaded than those of *M. tuberculosis*. The colonies are compact, domed and give a yellow-orange pigment.

Since the association of *M. scrofulaceum* with a case of lymphadenitis was first described in 1956, there have been many other cases reported from various parts of the world.

In adults, chronic pulmonary disease associated with *M. scrofulaceum* has been reported, but it is difficult to assess. Generally the organism colonizes a preformed lesion. Disseminated disease is almost invariably associated with some other serious disease.

M. szulgai

This scotochromogenic organism was first identified as a pathogen by Marks et al. (1972). When cultured at room temperature it exhibits photochromogenicity but at 37°C is scotochromogenic.

M. szulgai is widespread in its distribution: cases have been reported from Wales, Japan and the United States. The organism must be differentiated from *M. gordonae*, a slow growing scotochromogen that is frequently isolated from sputum of patients who do not have mycobacterial infection. This can be done using the Tween hydrolysis test: *M. gordonae* is positive and *M. szulgai* is generally negative. Urease and nitratase tests also aid identification of these organisms.

Runyon Group III. The Non-pigmented Mycobacteria

There is considerable overlap between the two major pathogens in this group, *M. avium* and *M. intracellulare*, making identification very difficult.

The organisms in Group III are too heterogeneous for successful identification by biochemical tests and lipid specificity. The most reliable method appears to be serotyping by agglutination and agglutinin absorption.

The organisms in this group are slow growing at room temperature. The colonies are mainly smooth and non-photochromogenic.

As *M. intracellulare* and *M. avium* are almost indistinguishable they are grouped together under one complex.

M. intracellulare and M. avium

Microscopically, the cells appear pleomorphic, but on routine culture they usually appear as short rods with bipolar acid-fast granules.

The optimal temperature is 40 °C, with growth occurring between 30 and 44 °C.

These mycobacteria are distributed throughout the world and have been isolated from a variety of sources, including water, soil, dairy products and the tissues of both birds and mammals. Avian bacilli produce spontaneous disease in domestic fowls and other birds. Man, swine and cows can become infected with these organisms.

The incidence of chronic pulmonary disease caused by *M. avium* and/or *M. intracellulare* is highest amongst middle-aged to elderly white males, with most of those affected having pre-existing pulmonary disease.

M. xenopi

This organism was isolated from a skin lesion on a South African toad, *Xenopus laevis,* and first described by Schwabacher in 1959. On a smear from a Lowenstein–Jensen culture the bacilli are generally long and thin and are often arranged in typical arching patterns.

The optimal temperature for growth is 42 °C. The colonies are formed after 3–4 weeks' incubation, giving a small flat colony sometimes showing slight pigmentation. Upon continuous incubation a yellow pigmentation often occurs.

M. xenopi has been isolated from water and granulomatous lesions in swine. In man it produces a chronic progressive pulmonary disease which is clinically and radiologically similar to tuberculosis. It has been seen more frequently in North-western Europe than in North America. In the United Kingdom it is associated primarily with people living near river estuaries.

M. xenopi is the only atypical mycobacterium which is relatively sensitive to isoniazid.

Runyon Group IV. The Rapid Growers

Rapid growers mature from small inocula at room temperature within 1 week on subculture. In this group there are at least two potential pathogens, *M. fortuitum* and *M. chelonei.*

M. fortuitum

These organisms are pleomorphic and exhibit various degrees of acid fastness. In pus, long and filamentous forms are seen, some showing definite branching. Colonies grown on Lowenstein–Jensen medium appear as rough, large colonies, often very similar to *M. tuberculosis,* but may take only 3 days to grow. It is therefore important to read cultures after 1 week to help distinguish between *M. fortuitum* and *M. tuberculosis. M. fortuitum,* unlike *M. tuberculosis,* can be cultured on ordinary agar media.

The most common clinical manifestation is an abscess appearing at an injection site of a supposedly sterile product. Cases of pulmonary disease are difficult to assess, pre-existing lesions generally being infected.

M. chelonei

This mycobacterium has undoubtedly been confused with *M. fortuitum* in the past. It was first reported by Friedmann in 1903 and described as the 'Turtle Tubercle Bacillus'. In recent years it has been recognized as a cause of abscesses and other infections in man and animals. Two varieties or subspecies have been described, the type subspecies occurring most commonly in Europe, and the abscessus subspecies more frequently in Africa and North America.

Colonies grow rapidly on ordinary media and Lowenstein–Jensen. However, some evidence suggests that primary isolation of the organism may take several weeks.

M. fortuitum and *M. chelonei* are antigenically quite distinct, with different phage patterns. Of the two species, *M. chelonei* now appears to be the more important pathogen of man.

Other rapid growers

There are many other rapid growing mycobacteria which very rarely, if ever, cause disease. *M. phlei* and *M. smegmatis* are commonly used in genetic and biochemical studies.

3 Safety Precautions

In recent years many committees and working parties have been set up to investigate the safety of staff working in pathology laboratories. Much of their work has been involved in the necessary precautions which should be observed in a tuberculosis laboratory.

The Health and Safety at Work Act (1974) and subsequent work have made it mandatory to observe various safety precautions, rather than being left to the discretion of each laboratory. *Mycobacterium tuberculosis* is a category B pathogen and as such the handling of all potentially infected material and cultures must be done in a Class I or III safety cabinet.

The function of the tuberculosis laboratory includes isolation, identification and sensitivity testing of many mycobacteria, some of which are resistant to the antituberculosis drugs available. It is therefore imperative that all staff within the laboratory appreciate the hazards involved in the handling of these specimens. Perhaps one of the greatest dangers of working in a tuberculosis laboratory is the familiarity with the work and the consequent laxness of technique and safety precautions. No measures can be too great to combat the risk of infection amongst laboratory workers.

AEROSOL FORMATION

Specimens, cultures and apparatus used are all potential sources of infection and one of the most important hazards is the formation of infected droplets or aerosols.

Aerosols are formed by the breaking of a fluid surface.

Larger droplets fall, whilst the smaller aerosols remain suspended in the air. The fluid may evaporate from these aerosols, leaving bacteria which are easily inhaled.

The opening of screw-capped bottles containing cultures on solid medium can create aerosols, as can centrifuging tightly stoppered or screw-capped bottles.

It is therefore essential to eliminate the risk of infection from these aerosols by observing certain safety precautions. It is desirable that all workers should have a positive tuberculin skin test, that regular chest X-rays be taken and that strict personal hygiene be enforced.

Many safety precautions are common sense but it is surprising how many times they have to be reiterated. The rule of no smoking, eating, etc. in the laboratory should be rigorously enforced. There should be no mouth pipetting in any laboratory, but especially in microbiology. Mechanical pipetting devices must be provided and must be used.

Facilities for hand washing should obviously be provided in the laboratory, with 'wrist type' taps for turning the water on and off. Disposable hand towels should be provided as roller-type towels too often fail to be changed regularly and can easily become infected.

Protective clothing should be worn in the laboratory when dealing with all infected or potentially infected material. These gowns should be confined to the TB laboratory and should be sterilized, by autoclaving, daily.

SAFETY CABINETS

All work involving the handling of specimens and cultures must be performed under cover of inoculating cabinets. Class I and Class III only should be used.

Class I Cabinet

This is an open-fronted exhaust protective cabinet which offers adequate protection to the worker against inhalation of aerosols containing category B organisms. Filtered air is discharged to the outside atmosphere by connected trunking. The air flow should be a minimum 0·75 linear metres per second (150 linear feet per minute) but should not exceed 1·0 linear metre per second (200 linear feet per minute).

Class III Cabinet

This is a totally enclosed exhaust protective cabinet which is gas tight and fitted with glove ports. This cabinet is used predominantly for Category A organisms and is not commonly found in hospital microbiology laboratories.

Decontamination

All exhaust protective cabinets must be washed down after use with a suitable disinfectant, e.g. 1% Hycolin (chlorinated phenol). Before cleaning the air grids, changing the filters or any maintenance work is carried out, the cabinets must be disinfected by using formaldehyde. There are two methods recommended by Howie:

1. Place 25 ml of formalin BP in a dish on an electric heater in the cabinet. Replace the front closure. If no manu-facturer's closure is provided one can be made of hard-board, or plywood, and attached to the working face with adhesive tape.
2. Place 35 ml of formalin BP in a 500 ml beaker in the cabinet. Add 10 g of potassium permanganate and replace the front closure. The mixture will boil in a few minutes, releasing formaldehyde.

   Caution: If too much potassium permanganate is added there is a risk of explosion. If too much formalin is used, polymer may be deposited in the cabinet and filters.

Leave the cabinet sealed overnight. The following morning switch on the fan and open the front closure to allow the air to enter and the remaining formaldehyde to be exhausted outside the building.

Testing the Air Flow of a Class I Cabinet

The air flow indicator must be checked daily. The air flow must be tested with a vane anemometer weekly if the cabinet is in daily use and monthly if used less frequently. Take five readings as shown in the diagram below.

X1		4X
	X3	
X2		5X

The air flow at each of these five positions must not be less than 0·75 linear metres per second (150 linear feet per minute) and should not exceed 1 linear metre per second.

Some cabinets may have an ultraviolet light inserted in the roof, as a means of sterilizing the cabinet. This is not an effective means of sterilization, as experiments have shown that even after 4 hours' exposure, bacilli in pathological material remain viable. Lamps may also continue to emit visible light after they have ceased to emit ultraviolet rays.

CENTRIFUGES

As previously stated, even screw-capped bottles in a centrifuge can discharge aerosols, thus making centrifuging a hazard. It is therefore essential that the centrifuges used for specimens containing potential Category B pathogens should have sealed buckets, and that these buckets are opened only in an exhaust protective cabinet.

The inside of the centrifuges should be washed with a suitable disinfectant daily and wherever possible the centrifuge bowl should be autoclaved once per week.

SHAKING MACHINES AND HOMOGENIZERS

Aerosols containing infected particles may escape from specimen containers on shakers and homogenizers. It is therefore ideal that these are housed in an exhaust protective cabinet.

Certainly homogenizers used for tissue specimens etc. *must* be operated in a cabinet.

DISINFECTANTS

There are numerous disinfectants available for laboratory use. It is always advisable to test the activity of the disinfectant chosen in the laboratory. Phenolic disinfectants, e.g. Hycolin, have been found to be very effective against tubercle bacilli. Ensure that the correct working solutions are prepared, and that they are changed daily.

Pipettes, slides and other disposable material should be discarded into jars containing a suitable disinfectant. It is advisable to autoclave the discard jars before throwing the contents away.

AUTOCLAVING

All material used in the tuberculosis laboratory should be sterilized by autoclaving. Cultures and sputum pots should be discarded into lidded containers and autoclaved.

4 Culture Media

INTRODUCTION

'Good culture media is the basis for a good microbiology laboratory.' This rather obvious statement is often forgotten today and many workers think of media making as a drudge and will either pass it on to whomever they can or buy commercially prepared media. Culture medium for the isolation of mycobacteria, however, is really very easily prepared, the main criterion being that it is as fresh as possible. Therefore only the amount required for a week or two should be prepared, not vast amounts which may look reliable but will reduce the isolation rate of mycobacteria because of its age.

The recommended culture medium in this country is still Lowenstein–Jensen. It is very reliable, easily prepared and can be used for primary isolation, sensitivity testing, identification and subculturing of the majority of mycobacteria.

The modified medium described by Jensen (1932) contained potato starch, but in a later publication by Jensen (1955) the starch was omitted. Many laboratories use Lowenstein–Jensen without potato starch and find this satisfactory, whilst others continue to add potato starch. Marks (1958) suggested that the inclusion of starch gave a slightly more luxuriant growth, and Baker (1967) found that the addition of starch reduced the amount of water of condensation.

Lowenstein–Jensen medium is very useful because any number of substances may be added to it prior to inspissation, such as the antimicrobial agents for sensitivity testing, *para-*

nitrobenzoic acid to distinguish *M. tuberculosis* and *M. bovis* from the atypical mycobacteria and sodium pyruvate to enhance the growth of the bovine strain. The pH of Lowenstein—Jensen medium is approximately 7·0. Acid Lowenstein—Jensen is used in pyrazinamide sensitivity testing and can be used as a primary isolation medium when no centrifuges are available (*see* pp. 32, 39).

The acidified Lowenstein—Jensen is softer and difficult to inoculate with a loop; a pasteur pipette is probably more reliable. The mycobacterial colonies grown on this medium are often much smaller and take longer to grow.

There are numerous liquid media available such as Youman's, Kirschner's and Middlebrook's 7H-9, and some workers still use these. They have the disadvantage of not giving colony counts and not being able to identify immediately whether it is growth of a mycobacterium or a contaminant.

A number of synthetic media have been described by Middlebrook and his colleagues following the medium described by Dubos and Davis (1946). The most successful and widely used of these media have been those of Dubos and Middlebrook, which have Tween 80 incorporated into the liquid media to give a dispersed type of growth as opposed to the 'clumpy' granular growth which occurs in Youman's medium. Middlebrook's 7H-10 and 7H-11 have agar added to them and are very useful media for studying colonial morphology. Although these media are complex they can be prepared in most laboratories with consistent results. Commercial preparations are available from Difco Laboratories.

Middlebrook's agar may be used as a primary isolation medium along with Lowenstein—Jensen but it is slightly more expensive, is more easily contaminated and does not appear to have a greater isolation rate. The role of synthetic media appears to be in colonial morphology studies and in sensitivity testing where the heating of an antimicrobial agent and protein binding are a great problem. Lowenstein—Jensen medium is easier to distribute than Middlebrook's agar medium which has to be kept molten whilst distributing.

PREPARATION OF CULTURE MEDIA
Lowenstein–Jensen Medium
Mineral Salt Solution

Potassium dihydrogen phosphate KH_2PO_4	4·0 g
Magnesium sulphate $MgSO_4$	0·4 g
Magnesium citrate	1·0 g
Asparagine	6·0 g
Glycerol	20·0 ml
Distilled water to	1000·0 ml

Dissolve by heating, distribute in 300 ml volumes
and autoclave at 121 °C for 20 minutes.

Malachite Green Solution

Malachite green	2·0 g
Distilled water	100·0 ml

Dissolve by heating, filter and autoclave at
121 °C for 15 minutes.

Complete Medium

Mineral salt solution	300 ml
Malachite green solution	10 ml
10 beaten eggs	

Method of Preparation

1. Clean the shells of 10 large eggs (approximately 500 g weight) with methylated spirit. Break the eggs aseptically into a sterile conical flask containing glass beads.
2. Shake the flask to break up the eggs, and add a 300 ml bottle of mineral salt solution.
3. Add 10 ml malachite green solution.
4. Shake well and filter through sterile butter muslin.
5. Distribute under aseptic conditions into sterile universal containers.
6. Inspissate at 75 °C until solidified. Keep the time constant for each batch of media.

N.B. If potato starch is to be included in the medium it may be added to the mineral salt solution (12 g to 300 ml) and heated over a flame until a smooth mixture is obtained. The solution is then steamed and used as described above.

Malachite green. It is imperative that the malachite green crystals purchased are chosen with care: many of those which are available are not suitable as they are bactericidal for mycobacteria. Ensure that those purchased have been tested for antimycobacterial activity, such as that obtainable from Merck.

Eggs. Fresh new laid chickens' eggs are the best for use in Lowenstein–Jensen medium; the fresher the eggs, the higher the isolation rate.

Penicillin Slopes

Some workers have found that incorporating 100 international units per millilitre of penicillin into the Lowenstein–Jensen medium reduces the contamination of the medium for primary isolation without inhibiting the growth of mycobacteria.

Prepare Lowenstein–Jensen medium as described and add 100 iu/ml of benzylpenicillin to the medium prior to inspissation.

Sodium Pyruvate Medium

The addition of sodium pyruvate to the Lowenstein–Jensen medium has been found to enhance the growth of *M. bovis* and certain drug-resistant strains of *M. tuberculosis* (Stonebrink, 1958; Marks, 1963). On an egg medium without sodium pyruvate *M. bovis* grows as a very fine, flat growth (dysgonic); with the addition of 0·5% sodium pyruvate a luxuriant (eugonic) growth is obtained. A final concentration of 0·5% sodium pyruvate is added to the Lowenstein–Jensen medium prior to inspissation in place of the glycerol.

PNB Medium

Para-Nitrobenzoic acid can be added to Lowenstein–Jensen medium as an aid to identification of mycobacteria. It was introduced by Tsukamura and Tsukamura (1964) who stated that when *p*-nitrobenzoic acid (sodium salt) was incorporated at a concentration of 500 μg/ml into Lowenstein–Jensen and an inoculum of 0·5–1 mg moist weight tubercle bacilli per ml was used only atypical mycobacteria would grow on it. Using the standard 4 mg/ml inoculum it has been found that a few strains of *M. tuberculosis* can grow on this medium, and the inoculum should therefore be diluted.

Method of Preparation

1. Weigh out 0·884 g of *p*-nitrobenzoic acid (equivalent to 1 g of the sodium salt) and add to 60 ml of distilled water containing 5·5 ml of N NaOH.
2. Heat to 60 °C and stir until dissolved.
3. Make volume up to 90 ml with distilled water and neutralize excess NaOH with N HCl (approximately 0·1–0·15 ml) using a pH meter or indicator paper.
4. Make up to 100 ml with distilled water to give a full concentration of 10 000 μg/ml.
5. This solution may be sterilized by membrane filtration or autoclaving and 5 ml added to every 100 ml of Lowenstein–Jensen medium prior to inspissation.

Youman's Medium (modification of Proskaur and Beck)

Asparagine	0·5 g
Potassium dihydrogen phosphate KH_2PO_4	0·05 g
Glycerol	2·0 ml
Distilled water	100·0 ml

Dissolve the ingredients in the water in the above order, ensuring that one is dissolved before the next is added. Adjust the pH to 7·0 with 4% sodium hydroxide and add magnesium citrate 0·15 g. Autoclave at 121 °C for 20 minutes. When cool, add sterile human, bovine or horse serum or plasma to give a final concentration of 10%.

Modified medium for pyrazinamide sensitivity testing: *see* p. 39.

Modified Kirschner's Medium (Ministry of Health, 1958)

Potassium dihydrogen phosphate KH_2PO_4	2·0 g
Disodium phosphate $Na_2HPO_4 . 12H_2O$	19·0 g
Magnesium sulphate	0·6 g
Sodium citrate	2·5 g
Asparagine	5·0 g
Glycerol	20·0 ml
Distilled water	1000·0 ml
Phenol red 0·4%	3·0 ml

Distribute in 9 ml volumes and autoclave for 10 minutes at 115 °C. For use, add 1 ml sterile horse serum. Penicillin 10 units/ml may be added to reduce contamination. Final pH should be 6·9–7·2. Phenol red indicates failure to neutralize material added and also detects contamination.

Mark's Medium for Niacin Testing

Disodium hydrogen phosphate Na_2HPO_4	7·5 g
Potassium dihydrogen phosphate KH_2PO_4	2·0 g
Magnesium sulphate $MgSO_4 . 7H_2O$	0·6 g
Sodium citrate	2·5 g
Iron and ammonium citrate	5·0 ml
Tryptone (oxoid)	5·0 g
Glycerol	20·0 ml
*Tween 80 (10% solution)	0·5 ml
*Sodium pyruvate	0·5 g
Phenol red 0·4%	0·3 ml
Distilled water to	1000·0 ml

*These are added to 100 ml of basic medium

Pyruvic acid should be neutralized with NaOH.
Autoclave the basic medium at 115 °C for 15 minutes. When cool add 5 ml of 8% Seitz filtered bovine albumin and 10 ml human citrated plasma. Final pH 7·4–7·6.

Middlebrook's 7H-9 Liquid Medium

Ammonium sulphate	0·5 g
L-Sodium glutamate	0·5 g
Sodium citrate, Na_3 citrate $2H_2O$	0·1 g
Pyridoxine hydrochloride	0·001 g
Biotin	0·0005 g

Disodium phosphate Na_2HPO_4		2·5 g
Potassium dihydrogen phosphate KH_2PO_4		1·0 g
Ferric ammonium citrate		0·04 g
Magnesium sulphate $MgSO_4 . 7H_2O$		0·05 g
Calcium chloride		0·0005 g
Zinc sulphate		0·001 g
Copper sulphate $CuSO_4 . 5H_2O$		0·001 g
Tween 80		0·5 ml
Distilled water	to	1000·0 ml

Dissolve in the water, bottle in 95 ml volumes and autoclave at 121 °C for 15 minutes. For use add 5 ml of bovine albumin-dextrose solutions and 0·3 ml catalase solution. (*See* p. 32.)

Middlebrook's 7H-10 Agar

Ammonium sulphate	0·5 g
L-Glutamic acid (sodium salt)	0·5 g
Sodium citrate Na_3 citrate $2H_2O$	0·4 g
Disodium phosphate Na_2HPO_4	1·5 g
Potassium dihydrogen phosphate KH_2PO_4	1·5 g
Glycerol	5·0 ml
Ferric ammonium citrate	0·04 g
Magnesium sulphate $MgSO_4 . 7H_2O$	0·05 g
Calcium chloride $CaCl_2 . 2H_2O$	0·0005 g
Zinc sulphate $ZnSO_4 . 7H_2O$	0·001 g
Copper sulphate $CuSO_4 . 5H_2O$	0·001 g
Pyridoxine hydrochloride	0·001 g
Biotin	0·0005 g
Malachite green	0·001 g
Agar powder (Difco)	15·0 g

The trace substances are conveniently prepared as stock solutions. First, dissolve the chemicals and then add and dissolve the agar powder; the medium can then be distributed in 90 ml volumes and autoclaved at 121 °C for 15 minutes. For use the agar is melted and cooled to 50 °C before adding 10 ml of oleic acid-albumin-dextrose-complex and 0·3 ml catalase solution. The poured plates should be kept away from daylight as much as possible as it has a deleterious effect on the medium and may prevent growth of mycobacteria.

Oleic Acid-albumin-dextrose Complex

Oleic acid		0·5 g
Bovine albumin fraction V		50·0 g
Dextrose		20·0 g
Sodium chloride		8·5 g
Distilled water	to	1000·0 ml

Dissolve the dextrose and bovine albumin in saline. Add the oleic acid to N/20 NaOH and then mix both solutions together. Filter through a Seitz filter and store at 4 °C. In 7H-10 medium used for capreomycin sensitivity testing, 10% sterile horse serum can be substituted for the oleic acid-dextrose complex.

Bovine Albumin-dextrose Solution

Bovine albumin fraction V		50·0 g
Dextrose		20·0 g
Distilled water	to	1000·0 ml

Dissolve in the water, filter through a membrane filter, bottle and store at 4 °C.

Catalase Solution

Crude catalase (beef liver)		1·0 g
Distilled water	to	1000·0 ml

Dissolve and filter through a membrane filter, bottle and store at 4 °C.

Acid Egg Medium (Kudoh Medium)

Potassium dihydrogen phosphate	10·0 g
Magnesium citrate	0·5 g
Magnesium glutamate	2·5 g
Glycerol	20·0 ml
Distilled water	500·0 ml
Eggs	1000·0 ml
2% Malachite green	6·4 ml

Mix all the salts into the distilled water and add to the mixed eggs. Add the malachite green. pH is 6·4.

Ogawa Egg Medium (Ogawa and Sanami, 1949)

This medium is used by Tsukamura for identification tests as an alternative to Lowenstein—Jensen medium.

Basal solution:

1% KH_2PO_4	100 ml
1% Sodium glutamate	100 ml
Whole eggs	200 ml
Glycerol	6 ml
2% Aqueous solution malachite green	6 ml

The ultimate pH is 6·8. The medium is distributed in 4 ml amounts made as slopes, and inspissated at 95 °C for 60 minutes.

Sauton Agar

Glycerol	30·0 ml
KH_2PO_4	0·5 g
$MgSO_4$	0·5 g
Citric acid	2·0 g
Ferric ammonium citrate	0·05 g
Sodium glutamate	4·0 g
Purified agar	20·0 g
Distilled water	970·0 ml

Adjust pH to 7·0 and dissolve the agar by heating at 100 °C. Distribute into 4 ml amounts and autoclave at 121 °C for 20 minutes. The medium is cooled and made as slopes.

Sula's Medium

This medium is very useful for the culturing of small numbers of tubercle bacilli which may be present in body fluids, e.g. pleural fluid, which should normally be sterile. The medium recommended by Ives and McCormick is double the strength of the original medium (Sula, 1947) so that an equal volume of body fluid can be added as an inoculum.

Disodium hydrogen phosphate	2·5 g
Potassium dihydrogen phosphate	1·5 g
Sodium citrate	1·5 g

Magnesium sulphate	0·5 g
Asparagine	2·0 g
Alanine	0·15 g
Glycerol	25·0 ml
Ferric ammonium citrate green scales	0·05 g
Malachite green (0·2% aqueous)	1·0 ml
Distilled water	500·0 ml
Ascitic fluid, sterile	50·0 ml
Penicillin solution 2000 units/ml	5·0 ml

Dissolve the salts, amino acids, glycerol and malachite green in water, distribute in 100 ml amounts in 10 oz bottles and autoclave at 121 °C for 15 minutes. Add the ascitic fluid with sterile precautions. Add the penicillin at time of inoculation with pleural fluid.

PREPARATION OF SENSITIVITY TEST MEDIA

Stock solutions of the various drugs are prepared fresh each time they are to be used. Where filtration (to sterilize) is required, membrane filtration and not Seitz filtration is used.

Preparation of the Lowenstein–Jensen slopes may be performed by one of two methods.

Method 1

Prepare the highest concentration of drug required in Lowenstein–Jensen medium and double dilute, using Lowenstein–Jensen medium as the diluent:

> *Example:*
> 100 ml of 64 μg/ml streptomycin in Lowenstein–Jensen medium
> +
> 100 ml of plain Lowenstein–Jensen medium
> = 200 ml of 32 μg/ml streptomycin in Lowenstein–Jensen medium

Repeat until the lowest concentration required is reached.

Method 2

Prepare × 50 the highest concentration of drug required, and double dilute with the required diluent, generally distilled water.

Add 1 ml of this dilution to each 50 ml of Lowenstein–Jensen medium.

Streptomycin Sulphate

Stock Solution

Dissolve 1 g in 5 ml of sterile distilled water.
Dilute a further 1/10 giving a 20 000 μg/ml solution.
Dilute a further 1/10 giving a 2000 μg/ml solution.

Method 1

Add 3·2 ml of 2000 μg/ml to 96·8 ml of Lowenstein–Jensen for each 100 ml of Lowenstein–Jensen sensitivity medium to contain 64 μg/ml streptomycin.
Add equal volumes of this to plain Lowenstein–Jensen medium to give 32 μg/ml streptomycin. Repeat down to 1·0 μg/ml.

Method 2

Add 4 ml of 20 000 μg/ml stock solution to 21 ml of sterile distilled water to give 3200 μg/ml. Double dilute using sterile distilled water down to 50 μg/ml. Add 1 ml of streptomycin solution to each 50 ml of Lowenstein–Jensen medium. Bottle and inspissate.

PAS

Stock Solution

Prepare stock solution to contain 2000 μg/ml in distilled water. Sterilize by membrane filtration.

Method 1

Dilute stock solution 1/10 and add 1 ml to 99 ml of Lowenstein–Jensen medium = 2 μg/ml. Double dilute in Lowenstein–Jensen medium.

Method 2

Dilute stock solution 1/20 and double dilute in sterile distilled water. Add 1 ml of each dilution to each 50 ml of Lowenstein–Jensen medium.

INH

Stock Solution

Prepare a stock solution to contain 10 000 μg/ml in distilled water. Sterilize by membrane filtration.

Method 1

Dilute stock solution 1/200 and add 4 ml to 96 ml of Lowenstein–Jensen medium = 2 μg/ml. Double dilute in Lowenstein–Jensen medium.

Method 2

Dilute stock solution 1/100 and double dilute in distilled water. Add 1 ml of each dilution to each 50 ml of Lowenstein–Jensen medium.

Rifampicin

Stock Solution

Dissolve 0·1 g in 10 ml of dimethyl formamide = 10 000 μg/ml.

Method

Dilute 1/5 and add 6·4 ml to 93·6 ml of Lowenstein–Jensen = 128 μg/ml. Double dilute in Lowenstein–Jensen medium.

Ethionamide

Stock Solution

Dissolve 0·2 g in 10 ml of absolute alcohol = 20 000 μg/ml.

Method 1

Dilute 1/5 in sterile distilled water and add 4 ml to 96 ml of Lowenstein–Jensen medium = 160 μg/ml.

Method 2

Add 8 ml of stock solution to 12 ml of distilled water = 8000 μg/ml. Double dilute in distilled water and add 1 ml to 50 ml of Lowenstein–Jensen medium.

Ethambutol

Stock Solution

Prepare stock solution to contain 10 000 μg/ml in distilled water. Sterilize by membrane filtration.

Method 1

Dilute stock solution 1/10 and add 1·6 ml to 98·4 ml of Lowenstein–Jensen = 16 μg/ml. Double dilute in Lowenstein–Jensen medium.

Method 2

Dilute stock solution 2/25 and double dilute in distilled water. Add 1 ml of each dilution to 50 ml of Lowenstein–Jensen medium.

Viomycin

Stock Solution

Add 10 ml of sterile distilled water to 1 g vial = 100 000 μg/ml.

Method 1

Dilute stock solution 1/20 and add 3·2 ml to 96·8 ml of Lowenstein–Jensen = 160 μg/ml. Double dilute in Lowenstein–Jensen medium.

Method 2

Dilute stock solution 2/25 and double dilute in sterile distilled water. Add 1 ml of each dilution to each 50 ml of Lowenstein–Jensen medium.

Cycloserine and Kanamycin

Stock Solution

Prepare stock solution to contain $100\,000\,\mu g/ml$ in sterile distilled water.

Method 1

Dilute stock solution 1/10 and add 3·2 ml to 96·8 ml of Lowenstein–Jensen $= 320\,\mu g/ml$. Double dilute in Lowenstein–Jensen medium.

Method 2

Dilute stock solution 4/25 and double dilute in sterile distilled water. Add 1 ml of each dilution to each 50 ml of Lowenstein–Jensen medium.

Thiosemicarbazone (Thiacetazone, TB1)

Stock Solution

Dissolve 0·1 g in 10 ml of dimethyl formamide $= 10\,000\,\mu g/ml$.

Method 1

Dilute stock solution 1/100 in sterile distilled water and add 4 ml to 96 ml of Lowenstein–Jensen $= 4\,\mu g/ml$. Double dilute in Lowenstein–Jensen medium.

Method 2

Prepare stock solution in tri-ethylene glycol: 1 g in 10 ml $= 100\,000\,\mu g/ml$. Dilute 1/10 in distilled water and take

1·6 ml of this plus 18·4 ml of 0·5% tri-ethylene glycol. Double dilute in 0·5% tri-ethylene glycol and add 0·5 l of each dilution to 100 ml of Lowenstein–Jensen medium.

Capreomycin

Stock Solution

Dissolve 1 g in 10 ml of sterile distilled water = 100 000 μg/ml.

Method

Dilute 1/25 and add 4 ml to 96 ml of Middlebrook 7H-10 medium. Melt and cool to 50 °C. Double dilute in 7H-10 medium.

Pyrazinamide

The amount of basic Lowenstein–Jensen medium used may vary according to the number of titrations required. The amount shown below is for 40 titrations.

Method

To 1600 ml of Lowenstein–Jensen medium, add 8 ml of 10 N HCl to give a pH 4·85. Weigh 0·5 g of pyrazinamide and add to 40 ml of distilled water to give 12 500 μg/ml. Add 32 ml of 12 500 μg/ml to 400 ml of Lowenstein–Jensen medium to give 1000 μg/ml. Double dilutions are then prepared from 1000 μg/ml to give the required dilution range.

'Stepped pH' Media for Pyrazinamide Sensitivity Testing
(Marks, 1964)

Liquid Medium

$Na_2 HPO_4$ anhydrous	0·75 g
$KH_2 PO_4$ anhydrous	0·2 g
$MgSO_4 7H_2 O$	0·06 g

Sodium citrate		0·25 g
Iron and ammonium citrate		0·5 mg
Tryptone (oxoid)		0·5 g
Glycerol		2·0 ml
Tween 80		0·05 ml
Pyruvic acid		0·2 ml
Phenol red 0·4%		0·3 ml
Distilled water	to	100·0 ml

Neutralize the pyruvic acid with NaOH. Autoclave the base at 10 lb for 15 minutes and when it is cool add 5 ml of 8% Seitz-filtered bovine serum albumin (Armour) and 10 ml of citrated human plasma from outdated transfusion blood. The final pH should be 7·4–7·6. Distribute 1·5 ml volumes in sterile bijou bottles containing a few glass beads and incubate overnight for sterility. Store at 4 °C for up to 4 weeks.

Egg Media

Any deviation from the procedure described may alter the buffer action of the medium and be detrimental to the test.

Prepare a 0·22% solution of pyrazinamide weighed and dissolved aseptically.

Basic egg medium. Prepare basic Lowenstein–Jensen medium as previously described. Add half the normal amount of malachite green.

Medium pH 5·2–5·4. Add 4·5 ml N HCl (Analar) to 100 ml of basic egg medium and mix vigorously without delay. Prepare drug medium similarly but with the addition of 2 ml pyrazinamide solution.

Medium pH 5·1–5·2. Prepare as the previous medium but use 5·0 ml N HCl.

Medium pH 5·0–5·1. Make 7·5 ml of autoclaved 9% KH_2PO_4 up to 100 ml with basic egg medium. Add 5·0 ml N HCl and mix at once. Prepare drug medium similarly but with the addition of 2 ml pyrazinamide solution.

Prepare medium for the fuller titration of the control strains by diluting part of the drug medium at each pH level 1 in 2 and 1 in 4 with the appropriate base medium.

Distribute the completed media in 2 ml volumes in bijou bottles. Lay these flat and inspissate.

Verify the pH of a set of inspissated media using methyl red. (British Drug Houses capillator method.)

5　The Treatment of Specimens

Handling of all specimens must be performed in a Class I safety cabinet.

SPUTUM

Sputum is by far the most common specimen received for tubercle examination in the laboratory. For the convenience of the patient a wide-necked sterile container is used. This allows the patient to produce the sputum into the container without contaminating the outside edges of the bottle. The sputum specimen is then transferred to a 1 oz glass universal container when it arrives in the laboratory. These bottles have the advantage of being stable, strong and able to withstand more than one centrifugation. Plastic universal containers do not appear to stand more than one centrifuging as they often split or leak the second time they are centrifuged.

Before concentrating the sputum, the amount and type of sputum should be recorded, stating the volume and whether it is saliva, mucoid, purulent or a combination of all three. The specimen is checked against the request form and given a laboratory number.

Decontaminating the sputum samples for the isolation of *M. tuberculosis* is necessary to kill any other bacteria which may be present such as normal flora, pathogenic bacteria or fungi. These other organisms grow more quickly than myco-bacteria and a large number of them are able to grow on Lowenstein–Jensen medium.

Transfer about 5 ml of the sputum sample to the 1 oz universal container. If the sputum is 'elastic' it may be necessary to cut the sputum with sterile scissors, which of course must be resterilized before re-using.

The value of preparing smears before processing a sample is a matter of opinion. A smear made at this stage is called a 'direct' smear, whilst a smear prepared after processing is

Table 2. Comparisons between Direct and Concentrate Smear Results

| | | Concentrate smear results | | | |
		Neg	+	+ +	+ + +
Direct smear results	Neg	73·9%	21·5%	2·6%	2·0%
	+	1·1%	39·2%	47·2%	12·5%
	+ +	—	2·7%	35·3%	62·0%
	+ + +	—	2·0%	6·0%	92·0%

called a 'concentrate' smear. In the experience of the author the latter is by far the better smear to detect positives because the sputum has been homogenized and any tubercle bacilli present have been evenly distributed throughout the specimen and are therefore more likely to be found in a smear.

Table 2 shows a comparison carried out at the Brompton Hospital between the direct and concentrate smears for acid-fast bacilli when stained by auramine-phenol. The total number of specimens counted was 1200. The results of the comparison showed, for example, that the number of smears which were + positive 'direct' smear and ++ positive 'concentrate' smear was 47·2%.

From a heavily infected sputum, growth will occur in about 2 weeks. The average length of time to achieve a good growth is from 3 to 5 weeks for pre-treatment samples. Tubercle bacilli which have been subjected to large amounts of chemotherapy, although still viable, will take much longer to grow. If incubator space allows, it is better to keep all cultures for up to 8 weeks before discarding as negative. Negative cultures giving a positive concentrate smear should be incubated for a further 4 weeks or longer, as growth often occurs after 8 weeks, particularly in patients with multiple resistant strains of *M. tuberculosis*.

Concentration Methods

Sodium hydroxide is toxic to tubercle bacilli and should be used at as low a concentration as possible without yielding too high a contamination rate. The percentage sodium hydroxide used in laboratories may vary as a result of climatic conditions and how heavily contaminated the specimen is to begin with, particularly in tropical zones where it may be several days before the sample can be cultured. Tubercle bacilli will survive in a sputum sample for many weeks, although the quicker the specimen is cultured the less likelihood there is of contamination occurring.

The alkaline concentration methods are the ones most commonly used nowadays. There are acid methods, using hydrochloric, sulphuric or oxalic acid, but although a very low contamination rate is achieved with these methods the isolation rate of tubercle bacilli is also much lower than with the alkaline methods. A good concentration method is one which finds a good balance between the contamination and isolation rate of tubercle bacilli. Acid methods kill approximately 90% of tubercle bacilli present in the sputum sample, with contamination rates of about 0·5%; alkali methods kill approximately 75% of tubercle bacilli present, with contamination rates of about 2%. If a method does not show *any* contamination, then generally speaking it is too harsh for the tubercle bacilli.

There are many concentration methods in use for the treatment of specimens for the isolation of *M. tuberculosis* but the following methods have proved over a number of years to be the most reliable and suitable for those laboratories dealing with small or large numbers of specimens.

Modified Petroff Technique

Petroff's method was introduced in 1915, and has, with modifications, proved to be one of the most efficient concentration methods. The method given is based on washing the deposit (after treatment with sodium hydroxide) with either sterile distilled water or sterile distilled water containing 100 iu/ml of benzylpenicillin. Penicillin water washing has been shown to reduce contamination. In the original Petroff method the washing was not done; a neutralization process using acid and a neutral red indicator was used. This method did not prove to be as reliable when dealing with large numbers of specimens and for this reason the washing of specimens is now normally done, although it does mean that each specimen has to be centrifuged twice.

Method

1. Add an equal volume of 4% sodium hydroxide to about 5 ml of sputum in a universal container (ensuring that the sodium hydroxide container does not come in contact with the neck of the specimen bottle).
2. Shake by hand or preferably place on a shaking or mixing machine for 15 minutes. Incubation at 37 °C during this time will not significantly raise the temperature of the contents unless a water bath is used, and this is not recommended as the water may become contaminated.
3. Centrifuge at 3000 r/min for 15 minutes taking the usual precautions such as balancing the containers and using sealed buckets (*see* p. 23).
   Do not open the centrifuge until it has stopped.
4. Pour off the supernatant fluid from the deposit into a suitable disinfectant (e.g. 2% Hycolin). Always use a double strength disinfectant to allow for the dilution of it by the supernatant fluid which is going to be tipped into it.
5. Using aseptic precautions add 15–20 ml of sterile distilled water or sterile water containing 100 iu/ml of

penicillin to the container. Replace the cap and shake for a few minutes to resuspend the deposit.
6. Centrifuge at 3000 r/min for 20 minutes.
7. Pour off the supernatant fluid into a disinfectant.
8. Inoculate from the deposit the entire surface of two Lowenstein–Jensen slopes using a Pasteur pipette. Smears are made from the residue left in the pipette.
9. Incubate the slopes in a horizontal position overnight at 37 °C and then for a further 8 weeks in an upright position.

Trisodium Phosphate Method (Corper and Stoner, 1946)

This method is less harsh than the sodium hydroxide method but does give a higher contamination rate. It is a method which is suitable for fresh specimens, not ones which have travelled from various chest clinics or have been flown from other countries.

Method

1. To about 5 ml of sputum in a universal container add an equal volume of 10% trisodium phosphate Na_3PO_4 and shake by hand to mix.
2. Incubate at 37 °C in a dry incubator for 24 hours.
3. Centrifuge at 3000 r/min for 30 minutes.
4. Resuspend deposit in sterile distilled water (or penicillin water, about 15–20 ml).
5. Centrifuge at 3000 r/min for 30 minutes.
6. Inoculate two Lowenstein–Jensen slopes as described in the Petroff method.

N-Acetyl-L-Cystine (NAC)

This mucolytic agent has been used with sodium hydroxide in the treatment of sputum for the isolation of mycobacteria. In comparisons of 4% NaOH and 2% NaOH plus NAC, Kubica and his colleagues (1963, 1964) showed an increase of about

30% in the number of positive cultures obtained using NAC. Lorian (1966) and Lorian and Lacasse (1967) using 2% NaOH with and without NAC found an increase of only 0·5–2% in positive cultures. In the United Kingdom the method using the 2% NaOH plus NAC tends not to be used due to the high contamination rate.

Acid Egg Method

This is a very good method for use in areas where there are no centrifuging facilities, and the sputum is heavily infected with tubercle bacilli.

Method

1. Mix an equal volume of 4% NaOH with the sputum.
2. Mix for 3 minutes and then using a sterile cotton wool swab inoculate two acid egg medium slopes (Kudoh medium).
3. Incubate for up to 8 weeks, reading the cultures weekly.

Microscopic Examination

Work done in recent years has shown that heat fixing of smears does not kill all tubercle bacilli present and could be hazardous to the staff handling and staining the smears. It is recommended therefore that the smears from the specimen are prepared, allowed to dry and fixed either in formalin vapour or a drop of mercuric chloride or alcohol, added to each slide.

The staining techniques generally used in the detection of mycobacteria are either the classical Ziehl–Neelsen method or, if fluorescence microscopy is used, the auramine–phenol or auramine–rhodamine technique (see pp. 61, 64).

Fluorescence microscopy has a number of advantages over the Ziehl–Neelsen method. In fluorescence microscopy, the microscopic examination is performed using the 4 mm and 16 mm objectives A much larger field can therefore be examined. The bacilli as tney are fluorescing appear larger than if stained by conventional methods, and it has been estimated that it is much quicker to examine a smear by fluorescence than to cover the same area with the Ziehl–Neelsen technique. Another factor is that in the Ziehl–Neelsen method oil immersion objectives are used, and it has been found that bacilli can float off one slide into the oil and then be transferred on to the next slide. This of course can lead to false positive reports unless the lens is carefully cleaned between each examination.

It is also far easier to see yellow fluorescing acid-fast bacilli on a black background than to find pink acid-fast bacilli on a blue or green background as in the Ziehl–Neelsen.

Slides should always be stained individually on a staining rack, with the slides at least 2 mm apart to prevent 'carry over' from one slide to the next as this can give false positive results.

Standard smears can be prepared for the grading of positive slides in different laboratories if standardization is required.

Preparation of Smears from Cultures

When making a smear from growth on Lowenstein–Jensen medium it is advisable to prepare the smear using a formalin or mercuric chloride solution or 70% alcohol. Place a drop of formalin, mercuric chloride or 70% alcohol on a slide and emulsify part of a colony in the solution, spreading to make a smear. Allow to dry and then heat fix. As previously mentioned the use of formalin, mercuric chloride or alcohol rather than distilled water is to ensure that the tubercle bacilli are properly fixed, as it has been shown that heat fixing of a culture smear is not sufficient to kill all the tubercle bacilli.

Preparation of Standard Smears (Mitchison, 1966)

In cooperative studies spread over a wide area, smears can be standardized and issued from one central laboratory so that an equivalent grading of results is obtained. It is convenient to have two smears, one in the heavy–moderate grade, the other in the moderate–scanty grade.

Method

A heavy positive sputum is diluted with an equal volume of water and shaken with glass beads for about 10 minutes to produce an homogenate.

Smears are made by delivering 0·005 ml of homogenate on to a glass slide and spreading over an area of 1 × 2 cm.

Fixation is carried out in an 80 °C oven with an atmosphere of formalin vapour.

For the first standard smear it may be necessary to dilute the sputum; this is done by adding non-tuberculous sputum to the homogenate.

Once a suitable sample has been prepared for the heavy–moderate standard a 1 : 15 dilution is prepared for the moderate–scanty standard.

Examination of Smears

The best results are obtained using fluorescence microscopy, but Ziehl–Neelsen staining may be used.

Method

1. Prepare an homogenate of the patient's sputum as described above or prepare smears in the usual manner.
2. Stain and examine by fluorescence microscopy.
   Smears with more bacilli than standard smear 1 = Heavy positive. Smears with less bacilli than standard 1 but more than standard 2 = Moderate positive. Smears with fewer bacilli than standard 2 but more than 2 bacilli seen = Scanty.

PROCESSING OF SPECIMENS OTHER THAN SPUTUM

Laryngeal Swabs

These swabs are often used on patients who have difficulty in producing a sputum sample, usually those from out-patient departments or chest clinics. The laryngeal swab is made using calcium alginate wool rather than cotton wool as this has been shown to give a higher recovery rate of tubercle bacilli.

Method

1. Immerse the swab in a 5 ml volume of 10% trisodium phosphate. After a few minutes the alginate wool dissolves into the fluid liberating the organisms on the swab. If cotton wool is used only the outer surface of the swab will be washed in the trisodium phosphate and many of the organisms will be trapped inside the swab.
2. Treat the specimen as for sputum, i.e. centrifuge etc.

Smears are not generally made owing to the small amount of material available and false negatives are likely to occur.

Gastric Contents

It is recommended that the aspirated contents are collected into a sterile 1 oz bottle containing 5 ml of 10% trisodium phosphate in order to neutralize the stomach acid which may begin to kill the tubercle bacilli present during transportation of the specimen to the laboratory. These specimens are useful in patients who do not produce very much sputum and are continually swallowing their saliva etc.

As soon as the specimen is received in the laboratory it should be centrifuged and the deposit treated as for sputum.

Urine Specimens

In the past a 24-hour collection was used, that is the total collection of urine passed by the patient in 24 hours. As the bottle was left on the ward all this time it became highly contaminated with other bacteria, and often when processed not all these organisms were killed and the Lowenstein—Jensen slopes became contaminated. Three consecutive early morning samples of urine are recommended nowadays. An early morning urine sample is collected and then immediately sent to the laboratory, thus reducing the chances of contamination.

Method

1. The specimen should be centrifuged at 3000 r/min for 30 minutes.
2. Remove the supernatant liquid and pool the deposits.
3. Treat as for sputum.

Pleural Fluid

Pleural effusions should be collected in a sterile 3·8% sodium citrate solution to prevent clot formation.

Method

1. Centrifuge the specimen and from the deposit inoculate two Lowenstein–Jensen slopes and prepare a smear for acid-fast bacilli.
2. Add 4% NaOH to the rest of the deposit and treat as for sputum.

Two sets of inoculations are carried out on this specimen to enhance the chances of isolation. The untreated specimen may contain only a few tubercle bacilli which might be killed by the sodium hydroxide. The sample is then treated to kill any other bacteria which might be present and which would contaminate the Lowenstein–Jensen medium.

Pus

Specimens of pus should be treated as a pleural fluid sample.

Cerebrospinal Fluid (CSF)

This specimen should always be treated with urgency. After collection, a sample of CSF from a case of tuberculous meningitis may produce a 'spider web' clot, although this does not appear to be very common nowadays. If a clot is found, a portion should be used to prepare smears which are stained for acid-fast bacilli as these organisms are often trapped in the structure of the clot.

As CSF is unlikely to contain more than one type of infecting organism, the deposit can usually be inoculated directly on to Lowenstein–Jensen medium; if, however, other organisms are seen, treatment should be as for pleural fluid.

Tubercle bacilli in smears prepared from CSF are often very scanty, and a long search should be made on a number of preparations. If there is sufficient fluid it should be centrifuged in a sterile tube before the smears are prepared from the deposit. It is recommended that a drop of the deposit is placed on a slide, not spread out but allowed to dry, and then another drop is placed on the same spot so that a thick smear is prepared.

Tissue

Biopsy specimens often contain only scanty numbers of tubercle bacilli and these may be the only organisms present. Cultures should therefore be prepared before and after treatment with NaOH.

Method

1. Place the tissue in a suitable container and with sterile instruments cut into small pieces.
2. Transfer pieces of tissue to sterile grinding tubes and grind to a paste. (It may be necessary to add sterile washed sand.)
3. Add sufficient sterile Dubos broth to mix and transfer the suspension to a fresh tube.
4. Prepare smears and inoculate one half of the specimen directly on to the Lowenstein–Jensen medium; treat the other half by the NaOH method, before inoculating a further two Lowenstein–Jensen slopes.

STORAGE OF CULTURES

Lowenstein–Jensen cultures should be kept for a suitable period (usually 6 months) after completion of sensitivity and identification tests in case further investigations are requested. The most satisfactory method is to freeze the cultures (filed in trays for easy reference) at $-20\,^{\circ}C$ in a deep-freeze. At this temperature viability is maintained for up to 2 years. For subculture the slope is removed from the deep-freeze and allowed to thaw at room temperature before transferring the growth to a fresh medium. Alternatively the organism can be grown in Dubos broth medium in a bijou container for 10 days, and then frozen either by deep-freeze or liquid nitrogen. Both these methods do not affect the morphology or biochemical reactions of the organism.

Cultures can also be freeze dried, but it is not recommended in the average laboratory as it is a very hazardous procedure when dealing with tubercle bacilli.

6 Sensitivity Testing of Mycobacteria

Today there are many antituberculosis drugs available and these are used by the clinicians in various combinations. Within a population of mycobacteria there are sensitive and resistant organisms. When a drug is given the sensitive organisms are killed and the resistant organisms are allowed to proliferate and develop greater resistance to the drugs. For this reason double or triple therapy is always given in the treatment of tuberculosis, the theory being that the resistant organisms will not be resistant to all the drugs given and complete eradication of all the bacteria will be obtained. If a patient does develop a resistant strain of *M. tuberculosis* during treatment this is termed 'secondary drug resistance'. However, if this patient then infects other people (who have never had therapy) with his resistant organism, that new patient is termed a 'primary drug-resistant patient'.

Sensitivity testing of mycobacteria is the estimation of the drug resistance of the organism and is used as an aid by the clinicians to select the appropriate drugs for chemotherapy and to change the drugs with the emergence of resistant strains. Sensitivity testing is therefore of vital importance and should only be performed in a laboratory which has the experience and facilities to perform large numbers of tests. Inaccurate sensitivity tests are probably more harmful to a patient than no sensitivity tests. Drug resistance is a problem which is encountered more and more in the United Kingdom today. At the present time in the United Kingdom single or multiple drug resistance occurs in about 2·5% of European patients and about 8% of immigrant patients (Yates, Collins and Grange, 1982). An accurate sensitivity test is therefore vital to ensure (1) that the patient is not given drugs to which the mycobacterium is resistant; or (2) that if the patient is on triple therapy (s)he is receiving at least two drugs to which the organism is sensitive. If only one effective drug were given, resistance to this drug could rapidly develop.

'First line drugs' is the name given to those which are the most effective and least toxic to the patient. The drugs included in this group are streptomycin, isoniazid, ethambutol and rifampicin. These are the first choice of the clinician. 'Second line drugs' is the name given to all the other anti-tuberculosis drugs.

In this country there are two main methods of sensitivity testing, the 'Resistance Ratio Method' and the 'Proportional Method'. A third method, the 'Absolute Concentration Method', is mainly used abroad. Although the proportional method is very good it is not easily adapted for routine and is used predominantly for research drug trials. In a diagnostic laboratory the resistance ratio method is far more practicable, provided the method is understood and carried out in detail.

RESISTANCE RATIO METHOD

This method is based on a comparison between the minimum inhibitory concentration (MIC)* of a known sensitive control strain of M. tuberculosis and that of the patient's strain. The known sensitive control strain is one which has been isolated from a patient who has never had any chemotherapy. The strain is a sensitive primary isolate, and is often referred to as a 'wild strain'. In the original method the known sensitive strain was the H37Rv, a strain of M. tuberculosis first isolated over 50 years ago from a patient and which became the universal control strain. Today many workers feel that this laboratory strain no longer bears any relationship with current mycobacterial isolates and should no longer be used as a control. Some workers, however, inoculate two or more control stains, one of which may be the H37Rv.

---

* The smallest amount of drug which inhibits the growth of an organism

Doubling concentrations of the drug are prepared in a solid medium (usually Lowenstein—Jensen) so that several identical titrations can be made and inoculated at one time. Sensitivity or resistance is determined by expressing the MIC of the test organism and the MIC of the control strain as a ratio.

Method

1. Using a 22 Standard wire gauge (s.w.g.) nichrome wire loop, with a 3 mm external diameter, a representative sweep from the primary culture is discharged into a bijou bottle containing six 3 mm glass beads and 0·2 ml sterile distilled water. The loopful of growth should be 4 mg moist weight of bacilli. Standardization of the inoculum size is most important and depends on estimating 4 mg of growth which has to be judged by eye. It is therefore advisable that one person prepare all of the suspensions, having gained experience by taking loopfuls of the culture and weighing them. This enables them to assess the weight of 4 mg.
2. Using a Vortex mixer, mix for 30 seconds.
3. Add 0·8 ml of sterile glass distilled water. This constitutes the suspension.
4. Using a 27 s.w.g. nichrome wire loop, 3 mm external diameter, inoculate the surface of each slope of the sensitivity test with a 'full' loopful of suspension.

A control drug-free slope, an agar slope and several identification slopes are set up for each strain to be tested. The control strains are also inoculated for each batch of tests, and again if the batch of medium is changed within a test. The identification slopes which are set up may vary from laboratory to laboratory; they should however be selected to give a good indication as to whether the strain is *M. tuberculosis* or an atypical mycobacterium.

Incubation of Tests

The slopes are incubated at 37 °C for 2 weeks. Some workers prefer to incubate for 3 or 4 weeks. In the experience of the author 2 weeks is perfectly adequate, and, provided that there is good luxuriant growth on the drug-free slopes, results can be read with confidence after 2 weeks' incubation. If the growth is poor on the drug-free slope then the tests and control slopes should be reincubated. If the slopes are incubated for too long a period the efficacy of the drugs starts to deteriorate, and then false sensitivity results may be given.

Reading and Interpretation

The MIC is taken as the lowest concentration of drug which shows less than 20 colonies on the slope.

The resistance ratio is obtained by dividing the MIC of the control strain into the MIC of the test organism.

A ratio of 1 : 1 or 2 : 1 indicates that the strain is sensitive.

A ratio of 4 : 1 is borderline, indicating that the strain is probably resistant, but the test should be repeated. If a 4 : 1 is obtained the second time then the strain should be considered resistant.

A ratio of 8 : 1 indicates a resistant strain.

These ratios apply for all drugs.

Drug Concentrations

Drugs are incorporated into Lowenstein–Jensen medium, before inspissation. Other solid media such as Middlebrook's 7H-10 or 7H-11 may be used, but Lowenstein–Jensen has proved to be very good and is less expensive than most other media. Drug concentrations used are as follows:

Streptomycin sulphate		1	2	4	8	16	32 μg/ml
PAS	0·03	0·06	0·12	0·25	0·5	1	2 μg/ml
INH 0·015	0·03	0·06	0·12	0·25	0·5	1	2 μg/ml
Ethionamide		5	10	20	40	80	160 μg/ml
Cycloserine		5	10	20	40	80	160 μg/ml
Thiosemi- carbazone (TB1)		0·12	0·25	0·5	1	2	4 μg/ml
Ethambutol		0·5	1	2	4	8	16 μg/ml
Rifampicin		2	4	8	16	32	64 μg/ml

Kanamycin	1	2	4	8	16	32	μg/ml
Pyrazinamide (acid Lowenstein–Jensen)	31	62	125	250	500	1000	μg/ml
Capreomycin (Middlebrook's 7H-10)	2	4	8	16	32	64	μg/ml

With capreomycin sensitivity testing Middlebrook's 7H-10 medium is used as the antibiotic becomes bound in an egg medium, reducing its activity. Pyrazinamide sensitivity testing is done in Lowenstein–Jensen medium but at an acid pH 4·85. Some workers use a liquid medium such as Youman's medium pH 4·4 (*see* p. 29).

Pyrazinamide sensitivity tests in Lowenstein–Jensen medium should be incubated at 37 °C for 4 weeks.

A 'stepped pH' technique for the estimation of pyrazinamide sensitivity has been described by Marks (1964) and is used by some laboratories. Using this method only one drug concentration is required and it caters for mycobacterial strains of different growth characteristics, especially pH susceptibility.

'Stepped pH' Technique for Pyrazinamide Sensitivity Testing

Method

Inoculate liquid medium (*see* p. 39) by emulsifying into it a heaped 1 mm loopful of a young primary culture. Incubate for 1 week at 37 °C. Before use, disperse the growth by mixing on a vortex mixer.

Egg medium. Withdraw a 1 mm loop edgewise from the shaken culture and inoculate an area about 5 mm in diameter at the exact centre of the slope. The loop should not touch shallow areas of the slope since the pH is easily raised and the growth resulting alters the reaction of the rest of the medium.

Reading and Interpretation

Tests are read after 3 weeks' incubation.

Each control with its corresponding drug tube is considered as a separate titration. It is common for strains to appear resistant at the higher pH levels and sensitive at the lower. In that event they are classed as sensitive. Resistance is only diagnosed when there is apparent growth in all readable titrations.

The MICs expected for sensitive control strains and H37Rv are as follows:

Streptomycin	2–4 μg/ml	Thiosemicarbazone	0·5 μg/ml
PAS	0·12–0·25 μg/ml	Ethambutol	1 μg/ml
INH	0·06 μg/ml	Rifampicin	2–4 μg/ml
Ethionamide	10–12 μg/ml	Kanamycin	2–4 μg/ml
Cycloserine	10–20 μg/ml	Pyrazinamide	31–62 μg/ml
		Capreomycin	4 μg/ml

These MICs are only a guide and may vary from laboratory to laboratory.

PROPORTIONAL METHOD

This method is based on the principle that although every sensitive strain of tubercle bacilli contains some resistant mutants, a resistant strain contains a greater proportion of them. Using drug-free Lowenstein–Jensen slopes and slopes containing very accurately weighed amounts of drug, viable counts are performed, at different concentrations of drug and at various dilutions of the patients organisms. The proportion of resistant organisms to sensitive ones can then be assessed. For full details the *Bulletin of the World Health Organisation* (W.H.O., 1963) should be consulted. It is the preparation of the drug-containing media and the very accurate standardization of the inoculum which makes this method a little impracticable in a routine laboratory.

ABSOLUTE CONCENTRATION METHOD

This method is rarely used in this country but is very popular abroad, it is a very simple test and only requires two drug-containing slopes for each drug tested plus a drug-free slope. Again the error in this method arises if the drugs are not very accurately weighed. Two concentrations are selected, e.g. isoniazid $0.2\,\mu g/ml$ and $1\,\mu g/ml$. If there is growth on the drug-free slope but none on the $0.2\,\mu g/ml$ and $1\,\mu g/ml$ slopes the organism is said to be sensitive, if there is growth on the drug-free and $0.2\,\mu g/ml$ slopes the organism is of intermediate susceptibility, and if there is growth on all three slopes then the organism is resistant. Controls of known sensitivity patterns are also set up at the same time.

The drug sensitivity patterns of atypical mycobacteria are quite different from *M. tuberculosis*. *M. tuberculosis* is normally sensitive to all the above antituberculosis drugs unless drug resistance has developed; with the atypical mycobacteria many are naturally resistant to the drugs. All atypical mycobacteria with the exception of *M. xenopi* are highly resistant to INH. *M. xenopi* has an MIC of about $0.25\,\mu g/ml$, making it relatively sensitive to INH. Runyon Group III mycobacteria, again with the exception of *M. xenopi*, are highly resistant to most antituberculosis drugs except for ethionamide and cycloserine. This makes them a very difficult group to treat, and often patients are subjected to 6-drug therapy or surgery as the only possible cure. Runyon Group IV mycobacteria are all very resistant to the antituberculosis drugs, again with the exception of ethionamide and cycloserine. Runyon Group I and Group II are more sensitive to the drugs than Groups III and IV but are generally resistant to at least two of the first line drugs.

7 Methods for Identifying Mycobacteria

There are many identification tests available for mycobacteria, but the first screening tests should be those which will rapidly distinguish *M. tuberculosis* and *M. bovis* from the atypical mycobacteria.

The other identification tests used should be those which will name the atypical mycobacteria associated with human disease or of medical importance down to species level. Several tests are required to identify the mycobacteria, there is no single test.

STAINING OF MYCOBACTERIA

Certain organisms, particularly the mycobacteria, have the ability of resisting decolorization by acids and alcohols when stained with certain dyes. This property of 'acid fastness' appears to be due to the large amount of lipoids (particularly mycolic acid) present on the cell wall of these organisms and causes not only acid fastness but a resistance to ordinary staining methods, such as Gram's stain.

Ziehl–Neelsen's Stain

Smears

The following solutions are required:

> *Carbol fuchsin*
> | | |
> |---|---|
> | Basic fuchsin | 1 g |
> | Absolute alcohol | 10 ml |
> | 5% phenol in distilled water | 100 ml |
>
> Dissolve the basic fuchsin in alcohol and then combine with the phenol solution.

Acid alcohol

Hydrochloric acid (s.g. 1·19)	3 ml
95% absolute alcohol (industrial methylated spirit)	97 ml

Methylene blue

Methylene blue	0·5 g
Distilled water	100 ml

Method

1. Fix smear.
2. Flood the slide with carbol fuchsin and heat (re-heat at intervals) and stain for 10 minutes.
3. Wash in running tap water.
4. Differentiate in 20% sulphuric acid or 3% acid alcohol.
5. Wash in running tap water.
6. Counterstain with methylene blue for 30 seconds. Wash.
7. Allow to dry and examine with the oil immersion objective.

Results

Acid-fast bacilli—red.
Cell nuclei—blue or green.

Sections

The following solutions are required:

Carbol fuchsin
1% acid alcohol
0·2% methylene blue or malachite green

Method

1. Deparaffinize with xylene, alcohol and water.
2. Flood the slide with carbol fuchsin and heat (re-heat at intervals) and stain for 10−15 minutes.
3. Wash with running tap water.

4. Differentiate with 1% acid alcohol until only the red blood corpuscles retain the stain when examined under the staining microscope.
5. Wash with running tap water.
6. Counterstain with 0·2% methylene blue or malachite green for 2 minutes.
7. Wash with running tap water.
8. Dehydrate rapidly, clear and mount in neutral balsam or DPX.

Results

Acid-fast bacilli—red.
Cell nuclei—blue.
Red blood corpuscles—pink.

Kinyoun's Stain

Smears

The following solutions are required:

Kinyoun's carbol fuchsin

Basic fuchsin	40 g
Phenol	80 g
Absolute ethyl alcohol	200 ml
Distilled water	1000 ml

Dissolve basic fuchsin in alcohol, add phenol, dissolve and then add water.
Acid alcohol 1%
Methylene blue (*see* p. 62)

Method

1. Fix smears.
2. Flood the slide with Kinyoun's carbol fuchsin and stain for 10 minutes.
3. Wash in running tap water.
4. Differentiate with acid alcohol until the smear is faintly pink.
5. Wash with water.

6. Counterstain with methylene blue for 30 seconds.
7. Wash and allow to dry.

Auramine-phenol Fluorescence Stain

Smears

The following solutions are required:

Auramine-phenol
3% aqueous phenol	100 ml

Warm to about 30 °C and add:

Auramine 0	0·3 g

Shake well and filter.

Method

1. Stain fixed smear with auramine-phenol for 10 minutes.
2. Wash with water.
3. Differentiate with 1% alcohol for 5 minutes.
4. Wash with water.
5. Flood slide with 0·1% potassium permanganate for 30 seconds to darken the background.
6. Wash and allow to dry in the air.
7. Examine by fluorescence microscopy.

Result

Acid-fast bacilli appear as bright luminous rods against a dark background.

Auramine-rhodamine Fluorescence Stain

The following solutions are required:

Auramine-rhodamine
Auramine 0	1·5 g
Rhodamine	0·75 g
Glycerol	75·0 ml
Phenol crystals (liquefied)	10·0 ml
Distilled water	50·0 ml

Mix glycerol and water, add phenol and shake thoroughly. Check solution is clear before dissolving in rhodamine. Finally, add auramine and shake thoroughly until dissolved. When preparing stain in bulk, it is advantageous to leave the mixture at 37 °C to ensure all the auramine is in solution. An oily scum which appears does not apparently interfere with staining but is best removed by filtration through No. 1 filter paper. This solution remains stable for several months at room temperature. It is advisable to filter it before use.

Method for Smears

1. Stain fixed smears for 10 minutes, heating until temperature has reached approximately 80 °C (i.e. until steam rises).
2. Decolorize for 2 minutes using 3% acid alcohol.
3. Rinse thoroughly in tap water and stain for 2–5 minutes with 0·1% potassium permanganate.
4. Wash and dry and examine by fluorescence microscopy.

Method for Sections

1. Deparaffinize with xylene, alcohol and water.
2. Stain with auramine-rhodamine stain for 10 minutes, heating to about 60 °C.
3. Wash with water.
4. Differentiate for 2 minutes in 1% acid alcohol.
5. Wash with water.
6. Counterstain with 0·1% potassium permanganate for 2 minutes.
7. Wash with water and leave to dry.
8. Dip in xylol and mount in DPX.
9. Examine by fluorescence microscopy.

Result

Acid-fast bacilli appear as golden-red rods against dark background.

IDENTIFICATION TESTS

All tests must be carried out in a Class I safety cabinet.

Niacin Test

There are three methods for niacin testing that have been found suitable for routine use. The first is a test made directly from the Lowenstein–Jensen culture using various chemicals, the second is a commercially prepared niacin test strip and the third is a test using a subculture of the organisms in a liquid medium.

Method 1

Reagents

1. Cyanogen bromide, 10% aqueous solution. This is a toxic substance and should be prepared in a fume cupboard and stored in a dark-brown glass stoppered bottle. The solution should keep for a month if stored in the cold.

2. 4% Aniline or 3% *o*-toluidine in 95% ethyl alcohol, stored in the cold in dark stoppered bottles. *o*-Toluidine is one of the substances listed as carcinogenic aromatic amines and the bottle should be labelled CARCINOGENIC —DANGEROUS. It is therefore better to use the 4% aniline.

Procedure

1. Using a mature culture (4 weeks old) grown on Lowenstein–Jensen medium at 37°C flood with 0·5 ml water for 30 minutes, allowing the water to soak over the slope.
2. Pipette the water into a test tube and add 0·5 ml of the 10% cyanogen bromide solution.
3. Add 0·5 ml 4% aniline in 95% ethyl alcohol, care being taken not to shake the tube.
4. A positive reaction is indicated by a bright yellow colour developing almost immediately.
5. An aliquot of 4% sodium hydroxide is finally added to the tube to prevent formation of HCN before discarding the tube.

An H37Rv culture is used as a positive control, 0·5 ml of distilled water as a negative control, as well as a known atypical mycobacterium.

Mechanism

The presence of niacin is indicated by the immediate production of a complex yellow-coloured compound; according to the Konig reaction a pyridine compound (niacin) reacts with cyanogen bromide and a primary amine (aniline) to produce a coloured product (Fernstein, 1945).

Results

M. tuberculosis	M. microti	M. bovis	Atypical mycobacteria
+	−	+ or −	−

Method 2. Niacin Test Strips

These are purchased commercially from Difco Laboratories and have proved to be very satisfactory. They are useful, as they prevent the need for using the toxic substances required in method 1. Bacto-TB niacin test strip is prepared with potassium thiocyanate, chloramine T, citric acid and sodium aminosalicylate.

Procedure

1. Add 1·5 ml of distilled water to a 3–4 week-old culture which has good growth. Stab a 1 ml sterile pipette through the growth into the medium to permit extraction of the niacin.
2. Slant the tube so that the surface of the medium is in the horizontal position and is covered with the liquid. Allow to soak for 20–30 minutes.
3. Carefully remove approximately 0·6 ml of the extract with a sterile capillary pipette and bulb and transfer to the bottom of a 13 × 75 mm test tube.

4. A negative control tube should be set up containing 0·6 ml of the same distilled water as used in extracting the culture. A known positive control organism should also be set up.
5. Drop a niacin test strip with arrow downward into each tube, and stopper immediately.
6. Shake tubes gently but do not tilt. Repeat this gentle shaking after 5–10 minutes.
7. After 12–15 minutes, but not later than 30 minutes, compare the colour of the extracts.

Results

A positive test for niacin is indicated by the appearance of a yellow colour in the extract of the test culture and no colour in the negative control.

Method 3 (Marks, 1965)

This method distinguishes between *M. tuberculosis* and other mycobacteria by the biological assay of niacin and it dispenses with cyanogen bromide. Results are obtained in about 8 days with eugonic and dysgonic strains.

Glassware should be scrupulously clean and kept for assay use only.

Medium. Oxoid niacin assay medium is the recommended culture medium and is prepared by dissolving 37·5 g in a litre of distilled water, bottled in 25 ml volumes and autoclaved at 115 °C for 10 minutes.

Test organism. Lactobacillus arabinosus 17-5 strain 8030 (obtainable from National Collection of Industrial Bacteria, Torry Research Station, Aberdeen, Scotland) is grown in glucose broth at 37 °C for 24 hours and then stored at 4 °C for up to 2 weeks.

Inoculum. A 2 mm loop is withdrawn edgewise from the culture and inoculated into 25 ml of assay medium.

Enrichment medium (see Marks medium, p. 30). Bottle in 1·5 ml volumes.

Niacin standards. Prepare a 0·1% solution of nicotinamide in distilled water and sterilize by autoclaving at 121 °C for 10 minutes. Dilute with sterile water to give standards containing 0·5 and 1·0 μg/ml. These solutions may be stored at 4 °C for many months. Nicotinamide is used in preference to niacin as it is more soluble.

Pipettes. A 20 μl pipette is used to transfer the culture.

Procedure

1. Inoculate a loopful of culture into a bottle of enrichment medium and incubate at 37 °C until definite growth occurs; this usually takes about 1 week.
2. Add 20 μl of test organism culture to 3 ml of inoculated assay medium and incubate at 37 °C for 24 hours. A blank and two standards are also tested.
   Blank—20 μl enrichment medium to 3 ml inoculated assay medium.
   Standards—20 μl of the 0·5 and 1·0 μg/ml solutions added to inoculated assay medium.
3. Add 9 ml distilled water to the tests and controls on completion and boil for 5 minutes to sterilize the culture.
4. Read the results in a nephelometer or absorptiometer.
5. A positive result is given when growth in the assay is equal to or greater than the highest control (1·0 μg/ml). Growth equal to or less than the lowest control is regarded as negative. If the result is between the two controls, the test is repeated.

Catalase-peroxidase Reaction

The catalase-peroxidase test is a valuable screening test for mycobacteria, catalase activity being shown by *M. tuberculosis* which is sensitive to isoniazid and by the atypical mycobacteria. The atypical mycobacteria generally show a much stronger effervescence. *M. tuberculosis* which is resistant to isoniazid loses both its catalase and peroxidase activity, whereas the atypical mycobacteria, even though they are highly resistant to isoniazid, always retain their catalase activity.

Reagents

1. 1% Hydrogen peroxide.
2. 0·2% Catechol in distilled water. (This substance is harmful by skin absorption, is corrosive, and should be handled with care.)

Method

1. Mix an equal volume of the two solutions, which should be prepared just prior to use, and add 5 ml to the test culture.
2. Allow to stand for a few minutes, observing any effervescence produced by catalase positive strains. Browning of the colonies indicates peroxidase activity.

Interpretation of Results

1. Catalase positive, peroxidase positive = *M. tuberculosis* sensitive to INH.
2. Catalase negative, peroxidase negative = *M. tuberculosis* resistant to INH.
3. Catalase positive, peroxidase negative = atypical mycobacteria.

Catalase Test (Kubica and Pool, 1960)

Only young cultures should be used as old cultures may give false negative results.

There are two methods:

1. Catalase activity of the culture tested directly on the medium.
2. Catalase heat-stable up to 68 °C which is tested in phosphate buffer pH 7·0.

Reagents

1. 10% Tween 80 in aqueous solution.
2. 30% Hydrogen peroxide.

These reagents must be freshly prepared and mixed in a 1 : 1 ratio.

3. Sorensen phosphate buffer pH 7·0.

Method

Catalase at room temperature

1. Add a few drops of Tween-hydrogen peroxide mixture to the Lowenstein–Jensen culture.

2. Th̶ ̶ ̶ bubbles of oxygen is noted.

̶velop immediately or after
Alternatively the growth can
̶e, placed in a test tube and

̶e in 0·5 ml Sorensen phos-
̶ered test tube.

̶ at 68 °C for 20 minutes
̶ bath must be placed in a

̶drogen peroxide mixture
̶ce of bubbles. The tubes
̶̶d of 20 minutes before

̶e for all mycobacteria
̶.. *tuberculosis* and *M. bovis* which are
̶̶tant to isoniazid.

Catalase heat-stable at 68 °C—positive for all atypical mycobacteria, negative for all strains of *M. tuberculosis.*

Arylsulphatase Test

This determines the presence or absence of the enzyme aryl-sulphatase. Arylsulphatase activity is found in most atypical mycobacteria but not in human or bovine strains of tubercle bacilli.

There are two methods that have been found satisfactory for general use, one using Dubos broth, the other using Dubos agar medium. Using the solid medium a 3-day arylsulphatase test is possible.

Method 1

Basal medium. Prepare Dubos broth base as manufacturer's instructions and then add 10% of a 5% bovine albumin solution which has been sterilized by membrane filtration and 1% of a 50% solution of glucose sterilized by autoclaving. This complete Dubos broth is then ready for use.

Weigh out 0·646 g of potassium phenolphthalein disulphate and dissolve in 100 ml distilled water to give a 0·01 M solution. Sterilize by membrane filtration and add 10 ml to each 100 ml of Dubos broth to give a final concentration of 0·001 M. Distribute in 5 ml amounts.

Procedure

Inoculate a 5 ml volume of the above medium with the organism to be tested and incubate for 14 days at 37 °C.

Test for arylsulphatase activity by adding 4% NaOH dropwise to the culture. A pink colour indicates a positive reaction.

Method 2 (Wayne, 1961)

Medium. To 100 ml of Dubos solid medium add 65 mg of potassium phenolphthalein disulphate (manufactured by Koch-Light Laboratories). After thorough mixing distribute the agar in 2 ml volumes in bijou bottles, autoclave at 121 °C for 15 minutes and allow to solidify in the upright position.

Procedure

Inoculate the medium with a portion of a colony using a straight wire to stab the surface of the medium.

Incubate at 37 °C for 3 days.

Add 0·5 ml 1·0 M sodium carbonate to the culture and observe for production of a pink colour indicating a positive result.

Mechanism

This test determines the presence or absence of the enzyme arylsulphatase which splits the bond between the sulphate group and the aromatic ring in the substrate by hydrolysis, thus liberating phenolphthalein which gives a red colour with alkaline sodium solutions. The intensity of the colour varies with the amount of enzyme produced. It is important that the NaOH and sodium carbonate are added to these tests with care as phenolphthalein is easily over-decolorized by a strong alkali. Controls should be set up with these tests: a positive control is *M. fortuitum* for both the 3-day and 14-day tests, a negative control for both is *M. tuberculosis*.

Results

	3 days	14 days
M. tuberculosis	—	—
M. bovis	—	—
M. kansasii	—	+ or —
M. fortuitum	+	+
M. xenopi	+	+
Other atypicals	—	+

Nitrate Reduction Test

Reduction of nitrate to nitrite is of value in distinguishing between human and bovine strains of tubercle bacilli. Human strains are generally strongly positive and bovine strains negative or weakly positive.

Reagents

1. M/15 phosphate buffer pH 7·0.
2. 0·02 M sodium nitrate in M/15 phosphate buffer pH 7·0 sterilized by membrane filtration.

3. Hydrochloric acid 1:1 dilution from concentrated with distilled water.
4. 0·2% Sulphanilamide in distilled water.
5. 0·1% N(1-naphthyl)-ethylenediamine dihydrochloride in an aqueous solution.

Store reagents 4 and 5 in the refrigerator and discard when they change colour. They remain stable for several weeks.

Method

1. Suspend approximately 10 mg (moist weight) bacilli in 1 ml phosphate buffer and add 1 ml of sodium nitrate solution. Mix well.
2. Incubate at 37 °C for 4 hours.
3. Add 1 drop of HCl solution.
4. Add two drops of sulphanilamide solution.
5. Add two drops of 1-naphthyl solution.
6. Mix and observe colour after 5 minutes: negative test remains colourless, positive shows intense red-violet colour.

A positive control, e.g. *M. tuberculosis,* and an uninoculated substrate as negative control should always be included.

A small amount of zinc dust should be dropped into all non-reactive tubes causing the reduction of nitrate to nitrite with the formation of a red colour, confirming the negative test.

Mechanism

$$NO_3 \xrightarrow[\text{Nitrate Reductase}]{\text{Nitrate}} NO_2$$

Nitrate · · · · · · · · · · · · Nitrite

If the mycobacteria contain nitrate reductase, nitrate is reduced to nitrite which undergoes a diazotization reaction with sulphanilamide which in turn couples with the naphthyl-ethylenediamine dihydrochloride, forming a pink compound.

Results

M. tuberculosis	+
M. bovis	—
M. kansasii	+
M. marinum	—
M. intracellulare	—
M. xenopi	—
M. fortuitum	+
M. chelonei	—

Tween 80 Hydrolysis Test (Wayne et al., 1964)

Medium. 100 ml M/15 phosphate buffer solution pH 7·0, 0·5 ml Tween 80 and 2·0 ml of 0·1% neutral red solution.

Distribute the medium in 4 ml amounts in screw-capped bottles and autoclave at 121 °C for 15 minutes. Store in the refrigerator.

Method

Inoculate one loopful of the test organism into the medium and incubate at 37 °C.

The colour is observed at 7 days and 14 days. The appearance of a reddish colour is regarded as positive.

A strain of *M. kansasii* is included as the positive control and for a negative control an uninoculated tube of Tween 80 substrate is used.

Tween 80. This is a detergent, a polyoxyethylene derivative of sorbitan mono-oleate. Tween 80 is in fact a complex mixture of polyoxyethylene esters of mixed partial oleic esters of sorbitol anhydrides with a 20% content of esterified oleic acid.

Mechanism

In the Tween 80 hydrolysis test the presence or absence of lipase is determined. The enzyme lipase hydrolyses the Tween 80 and forms free oleic acid. Neutral red is used as the indicator in the test, changing colour from amber (at pH 7·0) to pink-red at an acid pH produced by oleic acid formation.

Results

	7 days	*14 days*
M. tuberculosis	—	+
M. bovis	—	+
M. kansasii	+	+
M. scrofulaceum	—	—
M. intracellulare	—	—
M. xenopi	—	—
M. gastri, M. terrae	+	+
M. fortuitum	V	V

V = Variable

Tween Opacity Test (Wayne et al., 1964)

Medium. To molten Middlebrook's 7H-10 agar add 0·02% Tween 80 for use as a control and 2·5% for the test; dispense into bijou bottles and allow to set in an upright position.

Inoculum. One drop from a Pasteur pipette of the standard 4 mg/ml suspension is inoculated onto the surface of the agar. Incubate at 37 °C for up to 6 weeks.

Results

The culture grows only on the surface of the agar and if opacity is present it extends down the agar as incubation progresses.

Oleate Tolerance Test (Wayne et al., 1964)

This test is of value in the differentiation of the organisms in Runyon Group III. *M. gastri* gives a negative result and *M. intracellulare* and *M. terrae* are positive.

Medium. Middlebrook's 7H-10 agar with albumin and dextrose.

Control medium. Add 0·005% oleic acid.

Test medium. Add 0·025% oleic acid. Dispense as agar slopes in universal containers.

Inoculum. The standard sensitivity test inoculum is satisfactory.

Incubation. At 37 °C for up to 6 weeks.

Reading

A confluent growth (ignoring scanty colonies) is regarded as positive.

Strain	Catalase	Tween hydrolysis	Tween opacity	Oleate tolerance
M. terrae	High activity	+	+	+
M. gastri	Low activity	+	—	—
M. intracellulare	Low activity	—	+	+

Amidase Test

Bonicke and Lisboa (1959) described amidase tests which were subsequently used to aid identification of mycobacteria by the pattern obtained. A series of 10 amide solutions, called 'Bonicke's amide series', is inoculated with a mycobacterial suspension for 22 hours at 37 °C to observe the breakdown or not of the amides by the enzymes (amidases) to acids and ammonia (NH_3). *Table* 3 shows the chemical reactions of the 10 amides in Bonicke's series (*see below*).

The liberated ammonia is indicated by the use of phenol hypochlorite reagent which gives a blue compound with ammonia.

Reagents

1. *Bonicke's amide stock solutions.* The following quantities of amides are dissolved separately in 10 ml of distilled water (sterile) and heated to 80 °C in a waterbath for 30 minutes:

Amide No.	Amide	Weight (mg)
1	Acetamide	9·67
2	Benzamide	19·83
3	Urea (carbamide)	9·83
4	Isonicotinamide	19·99
5	Nicotinamide	19·99
6	Pyrazinamide	20·16
7	Salicylamide	22·45
8	Allantoin	25·89
9	Succinamide	19·00
10	Malonamide	16·71

Table 3. Chemical Reactions of Bonicke's Amide Series

Amide No.	Amide	Enzymatic Reaction	Name of Supplier
1	Acetamide	$CH_3CONH_2 \xrightarrow[\text{acetamidase}]{+H_2O} CH_3COOH + NH_3$ acetic acid ammonia	Merck, Darmstadt
2	Benzamide	$C_6H_5CONH_2 \xrightarrow[\text{benzamidase}]{+H_2O} C_6H_5COOH + NH_3$ benzamide benzoic acid ammonia	Merck, Darmstadt
3	Urea	$H_2N-CO-NH_2 \xrightarrow[\text{urease}]{+H_2O} CO_2 + 2NH_3$ urea carbon dioxide ammonia	Merck, Darmstadt
4	Isonicotinamide	isonicotinamide $\xrightarrow[\text{isonicotinamidase}]{+H_2O}$ isonicotinic acid $+ NH_3$ ammonia	C. Roth, Karlsruhe

Amide No.	Amide	Enzymatic reaction	Name of Supplier
5	Nicotinamide	nicotinamide $\xrightarrow[\text{nicotinamidase}]{+H_2O}$ nicotinic acid $+ NH_3$ ammonia	Merck, Darmstadt
6	Pyrazinamide	pyrazinamide $\xrightarrow[\text{pyrazinamidase}]{+H_2O}$ pyrazinoic acid $+ NH_3$ ammonia	Krugmann & Co., Hamburg
7	Salicylamide	salicylamide $\xrightarrow[\text{salicylamidase}]{+H_2O}$ salicylic acid $+ NH_3$ ammonia	C. Roth, Karlsruhe

(continued overleaf)

Table 3. Chemical Reactions of Bonicke's Amide Series (*continued*)

Amide No.	Amide	Enzymatic Reaction	Name of Supplier
8	Allantoin	NH_2 CO—CO—NH—CO NH—CH—NH allantoin $\xrightarrow[\text{allantoinase} \atop \text{allantoicase} \atop \text{urease}]{+4H_2O}$ $COOH$ C=O H glyoxylic acid + carbon dioxide $2CO_2$ + ammonia $4NH_3$	Fluka AG., Buchs SG., Switzerland
9	Succinamide	$CONH_2$ CH_2 CH_2 $CONH_2$ succinamide $\xrightarrow[\text{succinamidase}]{+2H_2O}$ $COOH$ CH_2 CH_2 $COOH$ succinic acid + $2NH_3$ ammonia	C. Roth, Karlsruhe
10	Malonamide	$CONH_2$ CH_2 $CONH_2$ malonamide $\xrightarrow[\text{malonamidase}]{+2H_2O}$ $COOH$ CH_2 $COOH$ malonic acid + $2NH_3$ ammonia	C. Roth, Karlsruhe

2. *Amide solutions 0·00164 M.* The stock solutions are diluted 1 : 9 with sterile distilled water.

3. *Manganese sulphate solution 0·003 M.* This is usable for several months if kept at 4 °C. 133·8 mg $MnSO_4 . 4H_2O$ is dissolved in 200 ml sterile distilled water.

4. *Phenol reagent.* This should be freshly made before use. 12·5 g phenol are added to 5 ml distilled water, shaken and added to 27 ml of a 5N sodium hydroxide solution (i.e. 20·0 g NaOH/100 ml distilled water). The solution is shaken to dissolve the phenol and made up to 50 ml with distilled water (i.e. 18 ml distilled water are finally added).

5. *Hypochlorite solution.* This is usable for several months if kept in brown bottles at 4 °C. Suspend 25 g of chlorinated lime in 300 ml distilled water and heat to 90 °C. Ensure that the lime is not too old and that chlorine can still be smelt. While shaking, add 135 ml 20% potassium carbonate (K_2CO_3) in aqueous solution and bring the mixture to the boil for 1 minute. Keep the temperature at 90 °C for 5 minutes and then leave to cool down. Fill the mixture up to 100 ml with distilled water (65 ml) and filter. *The first part of the filtrate is tested for free calcium ions* by adding a small amount of the potassium carbonate solution. If free calcium ions are still present, the solution becomes milky and more K_2CO_3 solution must be added to the mixture, which is then once more brought to the boil, kept at 90 °C for 5 minutes, left to cool down, filtered and stored.

6. *1/15 phosphate buffer, pH 7·2.* 11·876 g Na_2HPO_4 per litre and 9·078 g KH_2PO_4 per litre. 27·4 ml KH_2PO_4 solution are added to 72·6 ml Na_2HPO_4 solution. The solution is autoclaved at 121 °C for 15 minutes and dispensed into bottles in approximately 100 ml aliquots before sterilization.

7. *Reference series with ammonia.*

$10 \cdot 0 \, \mu g/ml$

$5 \cdot 0 \, \mu g/ml$

$3 \cdot 0 \, \mu g/ml$

$2 \cdot 0 \, \mu g/ml$

$1 \cdot 0 \, \mu g/ml$

The series is set up in the following way:

$1 \cdot 0$ ml of $2 \cdot 5\%$ ammonia soln. + $9 \cdot 0$ ml phosphate buffer, pH $7 \cdot 2$ = Soln. I

$1 \cdot 0$ ml of Soln. I + $9 \cdot 0$ ml phosphate buffer, pH $7 \cdot 2$ = Soln. II

$1 \cdot 0$ ml of Soln. II + $19 \cdot 2$ ml phosphate buffer, pH $7 \cdot 2$ = Soln.III = $10 \, \mu g/ml$

$1 \cdot 0$ ml of Soln. III + $5 \cdot 0$ ml phosphate buffer, pH $7 \cdot 2$ = $5 \cdot 0 \, \mu g/ml$

$1 \cdot 0$ ml of Soln. III + $7 \cdot 0$ ml phosphate buffer, pH $7 \cdot 2$ = $3 \cdot 0 \, \mu g/ml$

$1 \cdot 0$ ml of Soln. III + $8 \cdot 0$ ml phosphate buffer, pH $7 \cdot 2$ = $2 \cdot 0 \, \mu g/ml$

$1 \cdot 0$ ml of Soln. III + $9 \cdot 0$ ml phosphate buffer, pH $7 \cdot 2$ = $1 \cdot 0 \, \mu g/ml$

$1 \cdot 0$ ml of each solution containing 10, 5, 3, 2, and $1 \cdot 0 \, \mu g$ NH_3/ml is transferred to tubes and $0 \cdot 1$ ml $MnSO_4$ soln. + $1 \cdot 0$ ml phenol reagent + $0 \cdot 5$ ml hypochlorite soln. are added to each tube after which the tubes are heated to $100 \, °C$ for 15 minutes.

Procedure

The day before testing set up a series of tubes containing $0 \cdot 5$ ml of each of the amides and mark from 1 to 10 accordingly, *always* in the order that the amides appear on the list. The tubes are stored overnight at $4 \, °C$. The next day a fairly young, though well-grown, mycobacterial culture is used for making a 7 ml suspension in the phosphate buffer, corresponding to roughly 10 mg bacterial wet weight/$1 \cdot 0$ ml phosphate buffer. Pipette $0 \cdot 5$ ml bacterial suspension to each tube of the amide series from 1 to 10. Shake the tubes well and incubate at $37 \, °C$ for 22 hours. Remove from the incubator and to each tube of the series add the following reagents in this order:

$0 \cdot 1$ ml $MnSO_4$ soln.

$1 \cdot 0$ ml phenol reagent.

$0 \cdot 5$ ml hypochlorite soln.

Heat the tubes in a dry heat sterilizer to 100 °C for 15 minutes. After cooling compare the intensity of colour with the reference series.

Results

The results are graded according to the blue colour produced:

Results	Reference series µg NH_3/ml
Negative	0 and 1
Limit values	2 and 3
Positive	5
Strongly positive	10

Species	1	2	3	4	5	6	7	8	9	10
					Amidase Pattern					
M. tuberculosis	—	—	+	—	—	—	—	—	—	—
M. microti	—	—	+	—	+	+	—	—	—	—
M. bovis	—	—	+	—	—	—	—	—	—	—
Group I										
M. kansasii	—	—	+	—	+	—	—	—	—	—
M. marinum	—	—	+	—	+	+	—	—	—	—
Group II										
M. scrofulaceum	—	—	+	—	+	+	—	—	—	—
Group III										
M. intracellulare	—	—	—	—	+	+	—	—	—	—
M. avium	—	—	—	—	+	+	—	—	—	—
M. xenopi	—	—	—	—	+	+	—	—	—	—
M. gastri	—	—	+	—	+	—	—	—	—	—
M. terrae	—	—	—	—	V	V	—	—	—	—
Group IV										
M. smegmatis	+	+	+	+	+	+	—	(+)	+	(+)
M. phlei	(+)	—	+	—	+	+	—	—	—	—
M. fortuitum	+	—	+	—	(+)	(+)	—	+	—	—
M. chelonei	+	—	+	—	+	+	—	—	—	—

Sensitivity to Thiopen-2-Carboxylic Acid Hydrazide (TCH) and Furan-2-Carboxylic Acid Hydrazide (FCH)

M. bovis is sensitive to 1·0 µg/ml of both TCH and FCH while *M. tuberculosis* and the atypical mycobacteria are resistant to 25 µg/ml or more. However, since there is a cross-resistance between these two substances and INH, the test is therefore of no value in the case of isoniazid-resistant strains.

Method

Add TCH and FCH to Lowenstein–Jensen medium in a concentration of 1·0 µg/ml and 25 µg/ml prior to inspissation at 85 °C for 1 hour.

Inoculate using one 3 mm diameter loopful of a 4 mg/ml (moist weight) inoculum, incubate at 37 °C for 4 weeks and examine for growth.

Pigmentation

Some atypical mycobacteria form a pigment when grown in the dark whilst others may form a pigment only when exposed to light. Although this is not a particularly reliable test it can be of some value. Colours produced may vary from the usual buff colours of *M. tuberculosis* and *M. bovis* to yellow and deep orange with some of the atypical mycobacteria.

If a culture is being inoculated for pigmentation test it is advisable to pierce the cap of the container with a sterile hypodermic needle which has a cotton wool plug in the syringe attachment end. This allows a supply of air to reach the growing culture, and the development of pigmentation by some strains is greatly improved.

Two identical cultures should be tested for pigmentation, one in the dark and one exposed to light during the entire period of growth.

Growth at Different Temperatures

Inoculate four Lowenstein–Jensen slopes with the test organism and incubate at 25, 37, 44 and 52 °C. The growth is observed after 2 weeks and 4 weeks. The temperature at which abundant growth has occurred is recorded. Most atypical mycobacteria will grow at 25 °C, but *M. tuberculosis* and *M. bovis* do not.

ANIMAL INOCULATION

Animal inoculation is a test which is not very frequently used nowadays but there are certain circumstances when it may be necessary, e.g. the separation of a mixed culture of *M. tuberculosis* and *M. fortuitum* from a patient.

As *M. fortuitum* grows much more rapidly than *M. tuberculosis* and is resistant to most antibiotics it is almost impossible to separate the two species *in vitro*. Guinea-pig inoculation may be the only alternative when trying to establish whether or not the patient has a mixed infection.

All animal inoculations should be carried out in a safety cabinet. The animals should be housed in completely enclosed cages, not wire mesh types, and kept in a room specially designed for the purpose. After use the cages should be autoclaved together with feeding troughs, water bottles, etc.

Post-mortem examination of animals should be performed in a safety cabinet, the animal having first been soaked in a disinfectant to wet the fur. After the post-mortem the animal is placed in a plastic bag and burnt. Instruments should be sterilized by autoclaving and the safety cabinet disinfected.

The animal house should not be entered without first changing into rubber boots and gown. Before leaving, the gown should be discarded for sterilization and the boots washed in disinfectant.

Guinea-pig

A standard inoculum is prepared from the growth of a 2–3 week-old Lowenstein–Jensen culture.

Method

1. Weigh a sterile bijou bottle containing 12 3-mm diameter glass beads.

2. Add a loopful of culture to the dry bottle and reweigh.
3. Calculate the moist weight of bacilli and add water (or 0·1% bovine albumin) to give a final 2 mg/ml suspension. Shake the bottle for 1 minute to produce an even suspension.
4. Inject the animal in the thigh muscle with 0·5 ml (1·0 mg) of the prepared suspension.
5. Kill the animal after 6 weeks and examine for disease at the site of inoculation and for infected glands, liver, spleen and lungs.

Rabbit

A standard suspension is prepared from a 10—14-day culture in Middlebrook's 7H-9 Tween-albumin medium.

Method

1. A fully grown Tween-albumin culture contains approximately 0·6 mg moist weight bacilli per ml. Dilute this culture to a final strength of 0·002 mg/ml.
2. Inject the animal intravenously with 0·5 ml (0·001 mg) of the diluted culture.
3. Kill the rabbit after 6 weeks and examine for disease in the spleen, liver, lungs and kidneys.

Mouse

Method 1

Prepare a standard suspension containing 1·0 mg per ml moist weight bacilli. Inject the mouse in the tail vein with 0·2 ml (0·2 mg) using a 1·0 ml tuberculin type syringe with a 26 gauge needle.

Method 2

Inject a mouse intraperitoneally with 0·2 ml of a suspension containing 3 mg moist weight bacilli per ml.

With both methods the animals are killed after 6 weeks and examined by macroscopic and microscopic methods for evidence of disease.

Another method of producing disease in mice was described by Selbie and O'Grady (1954). Mice are injected with *M. tuberculosis* strain H37Rv in the thigh muscle and the lesion measured as it develops. This technique can be used to test antituberculosis agents in mice.

8 Identification of Mycobacterial Isolates

In the previous chapter a large number of identification tests were given to aid the identification of the mycobacteria. It is of primary importance that in a routine laboratory a group of screening tests be selected which will differentiate *M. tuberculosis* and *M. bovis* from other mycobacteria, rapidly and accurately. In a specialized tuberculosis laboratory all strains of mycobacterial isolates should be identified fully and named. It is imperative that these organisms be identified correctly, as the treatment of the patient may be determined by the laboratory identification, as well as the clinical significance of the isolate.

Cultures should be examined weekly and any culture showing growth is removed for further investigation. After 8 weeks' incubation, all cultures showing no growth, and which have a negative concentrate smear, are reported as 'negative cultures'. Those which give a positive concentrate smear are incubated for a further 4 weeks, or even longer if space will allow, as some mycobacteria which have been subjected to antituberculosis therapy take many weeks to grow.

SCREENING TESTS

On primary isolation a smear is made to test whether the organism isolated is an acid and alcohol-fast bacillus.

The most useful tests in a screening system are those which can be carried out quickly, accurately and with the minimum of reagents.

Recommended tests are:
1. Acid and alcohol-fast staining.
2. Colonial morphology on Lowenstein—Jensen medium.

3. Growth on nutrient agar.
4. Pigmentation.
5. Niacin production.
6. Catalase/peroxidase activity.
7. Growth on PNB medium.
8. Growth at different temperatures.
9. Sensitivity to INH and thiosemicarbazone.
10. Nitrate reduction.
11. Arylsulphatase test.

M. tuberculosis

1. Acid alcohol fast	—	Positive.
2. Colonial morphology	—	Rough, buff-coloured, breadcrumb-like colonies with corrugated surfaces.
3. Growth on nutrient agar	—	No growth occurs. Like most mycobacteria, *M. tuberculosis* requires egg- or serum-based media.
4. Pigmentation	—	No pigmentation occurs even when exposed to light.
5. Niacin	—	Positive.
6. Catalase	—	Positive when the organism is sensitive to INH, negative when the organism develops resistance to INH.
Peroxidase	—	Positive with INH sensitive organism and negative when the organism is INH resistant.
7. Growth on PNB	—	Negative.
8. Temperature	—	No growth at 25 °C or 45 °C. Optimal temperature for growth is 35–37 °C.
9. Sensitivity to INH and thio-semicarbazone	—	Generally sensitive to INH and Thiosemi-carbazone. Some strains are resistant to INH.
10. Nitrate reduction	—	Positive.
11. Arylsulphatase	—	3 days and 14 days negative.

M. bovis

1. Acid alcohol fast	—	Positive.
2. Colonial morphology	—	Dysgonic, smooth white colonies. Growth of *M. bovis* can be stimulated by the addition of sodium pyruvate to the medium.

3. Growth on nutrient agar — Negative.

4. Pigmentation — No pigmentation occurs, even when exposed to light.

5. Niacin — Generally negative.

6. Catalase — Weak positive.
   Peroxidase — Weak positive if organism is sensitive to INH.

7. Growth on PNB — Negative.

8. Temperature — No growth at 25 °C or 45 °C. Growth occurs at 37 °C.

9. Sensitivity to INH and thio-semicarbazone — Generally sensitive to both, but INH-resistant strains do occur.

10. Nitrate reduction — Negative.

11. Arylsulphatase — 3-day negative may be positive at 14 days.

M. africanum

1. Acid alcohol fast — Positive.

2. Colonial morphology — Eugonic, white colonies, intermediate between M. tuberculosis and M. bovis.

3. Growth on nutrient agar — Negative.

4. Pigmentation — No pigmentation occurs, even when exposed to light.

5. Niacin — Positive.

6. Catalase — Weak positive.
   Peroxidase — Not known.

7. Growth on PNB — Negative.

8. Temperature — No growth at 25 °C or 45 °C. Growth occurs at 37 °C.

9. Sensitivity to INH and thio-semicarbazone — Generally sensitive to both, but INH-resistant strains do occur.

10. Nitrate reduction — Variable.

11. Arylsulphatase — 3- and 14-day arylsulphatase negative.

Differentiation between M. tuberculosis and M. bovis is important for epidemiological reasons and can at times prove to be difficult. Microscopical examination of a stained culture smear does not help in the differentiation of these strains. On culture on Lowenstein–Jensen medium *M. bovis* gives a very dysgonic growth whilst *M. tuberculosis* gives a eugonic type of growth. The growth of *M. bovis* can be stimulated by the addition of sodium pyruvate to the medium. There is no growth variation when *M. tuberculosis* is inoculated on to a pyruvate-containing medium.

Niacin is produced by *M. tuberculosis* but *M. bovis* generally gives a negative reaction. Many workers today feel that the niacin test is not a useful test as it gives too many variables. However, it does, with the other screening tests, give an indication to identification.

Animal virulence is one of the oldest methods used in the differentiation of *M. bovis* from *M. tuberculosis*. The guinea-pig is susceptible to both of these organisms, providing they are INH sensitive. Variation has, however, been shown between various strains of INH-sensitive *M. tuberculosis*. Dhayagude and Shah in 1948 observed that strains isolated from Southern India were of low virulence for the guinea-pig compared to European isolates. Studies of virulence were made by Mitchison and his colleagues in the 1960s which resulted in a method being devised for grading virulence in the guinea-pig. The result of a virulence test gave a defined *root index of virulence* (Mitchison et al., 1960).

Rabbits, however, are only susceptible to the bovine strain and this animal can be used to differentiate the two strains. The animal should be inoculated intravenously with a diluted liquid culture and a post-mortem examination performed after 6 weeks. The bovine strain produces a generalized infection with lesions on the kidney, but the human type produces little disease and does not affect the kidney.

Growth on media containing thiopen-2-carboxylic acid hydrazide (TCH) aids the differentiation of the two species. It does, however, only separate INH-sensitive strains of *M. bovis* from *M. tuberculosis* and INH-resistant strains of *M. bovis*, and is therefore limited in its use.

Nitrates are generally reduced to nitrites by human strains but not by the bovine strains.

Table 4. Summary of Screening Tests for *M. tuberculosis* and *M. bovis*

Test	*M. tuberculosis* INH sens.	INH res.	*M. bovis*	Atypical mycobacteria
Acid alcohol fast	+	+	+	+
Colonial morphology	Rough	Rough	Dysgonic	Rough, smooth or dysgonic
Growth on nutrient agar	—	—	—	— or +*
Pigmentation	Buff	Buff	Buff	Buff, yellow or orange
Niacin production	+	+	—†	—
Catalase	+	—	+	+
Peroxidase	+	—	+ or —	—
Growth on PNP	—	—	—	+
Growth at 25 °C	—	—	—	+

*Runyon Group IV: the rapid growers are generally the only atypical mycobacteria to grow on nutrient agar.
† A weak niacin reaction may sometimes occur with *M. bovis* but generally a negative niacin test is obtained.

Table 5. Summary of Differentiation Tests for *M. tuberculosis* and *M. bovis*

	M. tuberculosis INH sens.	INH res.	*M. bovis* INH sens.	INH res.
Growth stimulated by pyruvate	—	—	+	+
Growth on TCH	+	+	—	+
Virulence for				
guinea-pig	+	±	+	±
rabbit	—	—	+	±
Niacin	+	+	—	—
Nitrate reduction	+	+	—	—

From these few tests *M. tuberculosis* and *M. bovis* can be differentiated from the atypical mycobacteria.

The results are summarized in *Tables* 4 and 5.

The Atypical Mycobacteria

Many workers have described schemes for the identification of mycobacteria with varying amounts of success; the scheme used here is the one used in the author's own laboratory.

One of the greatest problems in the isolation and identification of the atypical mycobacteria is to determine whether the organism isolated from a patient is of clinical significance. The criterion generally laid down is that the same atypical mycobacterium should be isolated from a patient's sputum on at least two occasions, preferably three, before being thought significant.

M. kansasii

1.	Acid alcohol fast	— Strongly positive.
2.	Colonial morphology	— Smooth, creamy white colonies. Some show rough variants which may make them indistinguishable from *M. tuberculosis* until they are exposed to the light.
3.	Growth on nutrient agar	— No growth.
4.	Pigmentation	— Produces a yellow pigment only when exposed to light.
5.	Niacin	— Negative.
6.	Catalase	— Vigorous catalase activity even though the organisms are resistant to INH.
	Peroxidase	— Negative.
7.	Growth on PNB	— This organism will grow on PNB medium.
8.	Temperature	— Will grow at 25 °C although it may take 3–4 weeks when compared with 2 weeks at 37 °C. No growth occurs at 45 °C.
9.	Sensitivity to INH and thiosemicarbazone	— Resistant to INH but generally sensitive to thiosemicarbazone.
10.	Nitrate reduction	— Positive.
11.	Arylsulphatase	— Negative at 3 days but 14 days shows variability.

The majority of *M. kansasii* strains will not grow on sodium pyruvate-containing medium, and it is therefore essential that this medium is not used solely for primary isolation.

M. marinum

1.	Acid alcohol fast	— Positive.
2.	Colonial morphology	— Soft, greyish white colonies, with slightly yellow streaks.
3.	Growth on nutrient agar	— None.
4.	Pigmentation	— Upon exposure to light at room temperature the colonies develop an intense orange-yellow pigmentation which eventually turns red.
5.	Niacin	— Negative.
6.	Catalase	— Weak catalase reaction.
	Peroxidase	— Negative.
7.	Growth on PNB	— Positive.
8.	Temperature	— Will grow at 25 °C slowly. Optimum temperature 31 °C.
9.	Sensitivity to INH and thio-semicarbazone	— Resistant to INH and thiosemicarbazone.
10.	Nitrate reduction	— Negative.
11.	Arylsulphatase	— 3 days negative, 14 days positive.

M. simiae

1.	Acid alcohol fast	— Positive.
2.	Colonial morphology	— Small dysgonic colonies, similar to M. intracellulare.
3.	Growth on nutrient agar	— No growth.
4.	Pigmentation	— Produces a yellow pigment very slowly when exposed to light.
5.	Niacin	— Positive.
6.	Catalase	— Positive.
	Peroxidase	— Negative.
7.	Growth on PNB	— Positive.
8.	Temperature	— Will grow at 25 °C and 37 °C. No growth at 45 °C.
9.	Sensitivity to INH and thio-semicarbazone	— Resistant to INH, but generally sensitive to thiosemicarbazone.

10. Nitrate reduction   — Negative.

11. Arylsulphatase   — Negative at 3 days but positive after 14 days.

M. scrofulaceum

1. Acid alcohol fast — Positive.
2. Colonial morphology — Eugonic, small dome-shaped colonies.
3. Growth on nutrient agar — Negative.
4. Pigmentation — Produces a yellow-orange pigment when grown in the light or dark.
5. Niacin — Negative.
6. Catalase — Positive.
   Peroxidase — Negative.
7. Growth on PNB — Positive.
8. Temperature — Will grow at 25 °C and 37 °C. No growth at 45 °C.
9. Sensitivity to INH and thio-semicarbazone — Resistant to INH and generally to thiosemi-carbazone.
10. Nitrate reduction — Negative.
11. Arylsulphatase — Negative at 3 days, positive after 14 days.

M. szulgai

1. Acid alcohol fast — Positive.
2. Colonial morphology — Rough colonies, eugonic growth.
3. Growth on nutrient agar — Negative.
4. Pigmentation — When cultured at room temperature it produces a pigment only after exposure to light. At 37 °C the organism is scotochromo-genic, giving yellow pigmented colonies, in the dark.
5. Niacin — Negative.
6. Catalase — Positive.
   Peroxidase — Negative.

7. Growth on PNB — Positive.

8. Temperature — Will grow at 25 °C and 37 °C. No growth at 45 °C.

9. Sensitivity to INH and thiosemicarbazone — Resistant to INH and thiosemicarbazone.

10. Nitrate reduction — Positive.

11. Arylsulphatase — Negative at 3 days, positive after 14 days.

M. gordonae

1. Acid alcohol fast — Positive.

2. Colonial morphology — Small, smooth colonies.

3. Growth on nutrient agar — Negative.

4. Pigmentation — Produces yellow-orange pigment when grown in the dark.

5. Niacin — Negative.

6. Catalase — Positive.
   Peroxidase — Negative.

7. Growth on PNB — Positive.

8. Temperature — Will grow at 25 °C and 37 °C. No growth at 45 °C.

9. Sensitivity to INH and thiosemicarbazone — Resistant to INH and thiosemicarbazone.

10. Nitrate reduction — Negative.

11. Arylsulphatase — Negative at 3 days, positive after 14 days.

M. avium, M. intracellulare

1. Acid alcohol fast — Positive.

2. Colonial morphology — Smooth, dysgonic, opaque colonies.

3. Growth on nutrient agar — Negative.

4. Pigmentation — Generally non-pigmented.

5. Niacin — Negative.

6. Catalase	— Positive.
Peroxidase	— Negative.
7. Growth on PNB	— Positive.
8. Temperature	— Will grow at 25 °C, 37 °C and 45 °C. Grows very slowly at 25 °C.
9. Sensitivity to INH and thiosemicarbazone	— Highly resistant to INH and thiosemicarbazone.
10. Nitrate reduction	— Negative.
11. Arylsulphatase	— Negative at 3 days. *M. intracellulare* positive at 14 days. *M. avium* negative at 14 days.

M. xenopi

1. Acid alcohol fast	— Positive.
2. Colonial morphology	— Small, smooth colonies.
3. Growth on nutrient agar	— Negative.
4. Pigmentation	— Sometimes shows slight pigmentation. On prolonged incubation a yellow pigmentation often occurs.
5. Niacin	— Negative.
6. Catalase	— Positive, but only a weak reaction may occur.
7. Growth on PNB	— Positive.
8. Temperature	— Will not grow at 25 °C, grows at 37 °C and 45 °C. Optimal temperature for growth is 42 °C.
9. Sensitivity to INH and thiosemicarbazone	— Relatively sensitive to INH but highly resistant to thiosemicarbazone.
10. Nitrate reduction	— Negative.
11. Arylsulphatase	— 3 days and 14 days positive.

M. ulcerans

1. Acid alcohol fast	— Positive.
2. Colonial morphology	— Small, domed, smooth colonies.

3. Growth on nutrient agar — No growth.

4. Pigmentation — Produces a pale cream to yellow colony. Shows slight pigmentation.

5. Niacin — Negative.

6. Catalase — Positive.
   Peroxidase — Negative.

7. Growth on PNB — Negative.

8. Temperature — Optimal temperature for growth is 30 °C.

9. Sensitivity to INH and thiosemicarbazone — Resistant to both.

10. Nitrate reduction — Negative.

11. Arylsulphatase — 3 days and 14 days arylsulphatase negative.

M. fortuitum

1. Acid alcohol fast — Positive.

2. Colonial morphology — Rough, large colonies.

3. Growth on nutrient agar — Positive, will grow in 3 days.

4. Pigmentation — No pigmentation occurs.

5. Niacin — Negative.

6. Catalase — Positive.
   Peroxidase — Negative.

7. Growth on PNB — Positive.

8. Temperature — Will grow at 25 °C and 37 °C. No growth at 45 °C.

9. Sensitivity to INH and thiosemicarbazone — Highly resistant to INH and thiosemicarbazone.

10. Nitrate reduction — Positive.

11. Arylsulphatase — 3 days and 14 days positive.

M. chelonei

1. Acid alcohol fast — Positive.

2. Colonial morphology — Rough colonies produced by subspecies chelonei. Smoother colonies produced by subspecies abscessus.

Table 6. Summary of Screening Tests for Mycobacteria

	Niacin	Catalase	Growth Rate	Growth on N.A.	PNB	25°C	37°C	45°C	52°C	Nitrate	Aryl-sulphatase 3-day	14-day	Pigmentation Light	Dark
M. tuberculosis INH sensitive	+	+	S	−	−	−	+	−	−	+	−	−	−	−
M. tuberculosis INH resistant	+	−	S	−	−	−	+	−	−	+	−	−	−	−
M. bovis	−	−	S	−	−	−	+	−	−	−	−	−	−	−
M. ulcerans	−	+	S	−	−	+	−	−	−	−	N∓	N∓	±	±
M. kansasii	−	+	S	−	+	+	+	−	−	+	−	±	+	+
M. marinum	−	±	S	−	+	+	−	−	−	−	−	+	+	+
M. simiae	+	+	S	−	+	+	+	−	−	−	−	+	+	+
M. scro-fulaceum	−	+	S	−	+	+	+	−	−	−	−	+	+	+
M. szulgai	−	+	S	−	+	+	+	−	−	+	−	+	+ at 37°C	+ at 37°C
M. gordonae	−	+	S	−	−	+	+	−	−	−	−	+	+	+
M. avium	−	+	S	−	−	±	+	+	−	−	−	−	−	−
M. intra-cellulare	−	+	S	−	−	±	+	+	−	−	−	−	−	−
M. xenopi	−	+	S	−	+	−	+	+	−	−	±	+	±	±
M. gastri	−	−	S	−	+	+	+	−	−	−	−	+	−	−
M. terrae	−	+	S	−	+	+	+	−	−	+	−	−	−	−
M. fortuitum	−	+	R	+	+	+	+	−	−	+	+	+	−	−
M. chelonei	−	+	R	+	+	+	+	−	−	−	+	+	−	−

N.A. = nutrient agar; S = slow; R = rapid.

3. Growth on nutrient agar — Positive. Will grow after primary isolation in 3 days.
4. Pigmentation — Non-pigmented.
5. Niacin — Negative.
6. Catalase — Positive.
   Peroxidase — Negative.
7. Growth on PNB — Positive.
8. Temperature — Will grow at 25 °C and 37 °C. No growth at 45 °C.
9. Sensitivity to INH and thio-semicarbazone — Highly resistant to INH and thiosemi-carbazone.
10. Nitrate reduction — Negative.
11. Arylsulphatase — 3 days and 14 days positive.

Table 6 summarizes the results of the various screening tests for the majority of mycobacteria encountered in a routine diagnostic laboratory.

9 Antituberculosis Drugs

ASSAY METHODS

As previously mentioned (p. 55), the first line drugs for the treatment of tuberculosis are INH, streptomycin, rifampicin and ethambutol. The second line drugs include *para*-amino salicylic acid (PAS), ethionamide, thiacetazone, and pyrazinamide. Although assay methods are not commonly used for antituberculosis drugs they are of value in certain circumstances. Assay methods are used to determine the amount of drug present in the serum or urine of patients. These methods may be used to check that the patient is actually taking the drug, in which case a screening test is adequate, or to determine the actual amount of drug in the patient's serum at a given time after injection. This is performed to ensure that there is not an accumulation of drug in the patient due to some malfunction, e.g. renal failure.

Isoniazid

Isoniazid (iso-nicotinic acid hydrazide, INH) is absorbed along the whole length of the gastrointestinal tract, and there is very little evidence to suggest that its excellent absorption is ever impaired. It penetrates readily through cell membranes and passes rapidly into the CSF, even in the absence of inflammation.

Isoniazid is generally metabolized to acetylisoniazid, and only a small proportion of the dose is excreted unchanged in the urine.

Determination of Isoniazid in Fresh Plasma

Method (slightly modified from Dymond and Russell, 1970)

Pipette 1 ml plasma into a stoppered centrifuge tube together with 1 ml M dipotassium hydrogen phosphate and 1·5 ml of freshly prepared aqueous 0·5% 2, 4, 6 trinitrobenzene-sulphonic acid. After mixing add 3 ml methyl-isobutyl-ketone and stand in the dark for 15 minutes. Shake thoroughly, separate the phases by centrifugation and measure the optical density of the upper phase at 480 nm. The concentration of isoniazid can then be calculated from the results obtained from standards containing 0–10 mg/l isoniazid in blank plasma.

There are other methods for the determination of iso-niazid, including colorimetric, fluorimetric and microbio-logical methods.

Determination of Isoniazid in Urine (Screening Test, method of Eidus and Hamilton, 1964)

Reagents

Potassium cyanide, 10% solution
Chloramine T, 10% solution
Store at 4 °C for 1 week.

Method

On to a white porcelain tile (with depressions) place 4 drops of urine, 4 drops of potassium cyanide and 10 drops of chlor-amine T solution. The presence of acetylisoniazid is shown by the development of a pink-red colour.

Streptomycin

Streptomycin is administered by intramuscular injection as there is little absorption from the gastrointestinal tract. It is important that streptomycin is monitored, as excess can cause deafness, giddiness and renal impairment.

The microbiological method of Mitchison and Spicer (1949) is still the one of choice.

Streptomycin Assay by Agar Diffusion Method

Required

1. 19 ml of assay agar (equal volumes of nutrient agar and 1% peptone in distilled water at pH 8·0).
2. Glass tubing, 3 mm internal diameter and 8 cm in length, sealed at one end and sterilized in the hot-air oven.
3. 18-hour broth culture of 'Oxford' *Staphylococcus aureus*.
4. Streptomycin standards in serum containing 4 mg/l and 50 mg/l.
   Test serum.

Method

1. Add to the agar (melted and cooled to 45 °C) 1 ml of a 1/50 dilution of the 18-hour culture of Oxford *Staph. aureus* (final dilution 1/1000).
2. To 18 tubes add, rapidly, sufficient seeded agar to give a column of 4 cm.
3. Ensure that the meniscus is even.
4. To 6 tubes add sufficient of the standard 4 mg/l streptomycin to give a layer of 1 mm or more serum on and above the agar.
5. Repeat to the next 6 tubes with the 50 mg/l streptomycin standard.
6. Repeat to the next tubes with the test serum.
   N.B. Many sera may be tested at the same time, provided only 1 container of seeded agar is used for each batch.
7. Incubate at 37 °C for 24 hours.

Reading of Results

The streptomycin in the serum will diffuse into the medium, inhibiting the growth of the *Staph. aureus*. The length of this zone of inhibition will depend on the amount of streptomycin in the serum. Measure the zone of inhibition using a microscope with a 16 mm objective, an eyepiece with a crosswire and a mechanical stage with a Vernier millimetre scale.

1. Place a 75 × 25 mm microscope slide in position under the microscope.
2. Fix a small ridge of Plasticine at both ends of the slide and place one of the tubes across the ridges.
3. Focus onto the agar meniscus so that the crosswire of the eyepiece is directly across the meniscus.
4. Read the Vernier scale.
5. Traverse the stage until the crosswire is directly across the edge where colony growth commences.
6. Read the Vernier scale.
   Subtraction of this reading from the first reading gives the length of the zone of inhibition in millimetres.
7. Take an average of the readings for the two controls and the test sera.
8. Square these figures, and plot on a graph as follows:
   Along one side of the graph plot the logarithm to the base 10 of streptomycin concentration. Up the other side of the graph plot the squares of the zone of inhibition. Plot the two control points on the graph and draw a straight line through them. Use this line for reading off the unknown sera.

Rifampicin

Rifampicin is absorbed well after oral administration and is excreted in the urine and bile. It appears to penetrate readily through cell membranes and passes rapidly into the CSF, even in the absence of inflammation.

There are many methods for rifampicin estimation; the one given below is relatively simple and can be used in a routine laboratory.

Method

1. Pour 100 ml DST agar (Oxoid) into a 24 cm square assay dish (NUNC). Allow to set on a flat surface.
2. Using the overnight broth culture, diluted 1/50, of either the Oxford *Staph. aureus* (NCTC 6571) or

streptomycin-resistant strain of *Staph. aureus* (NCTC 20702), seed the plate.
3. Allow to dry.
4. Using a 4·5 mm diameter punch, prepare sufficient wells to test the control sera and test sera in triplicate.
5. Prepare standard solutions of rifampicin in serum. Required standards: 10·0, 5, 2·5, 1·2 and 0·625 mg/l.
6. Using a Pasteur pipette fill each well with the appropriate serum, each sample and each test being done in triplicate.
7. Incubate at 37 °C overnight.

Reading and Interpretation

Plot mean zone diameter for each standard versus the log of the concentration using semi-logarithmic graph paper. Read test results from the straight line.

Estimation of Ethambutol (Lee and Benet, 1976)

Method
1. Add 5 μg *dextro*-2-(ethylenediamino)-di-1-propanol as an internal standard to 0·2 ml of plasma, and extract for 10 minutes under alkaline conditions with 8 ml of chloroform.
2. Transfer portions of the chloroform extract to another tube and evaporate to dryness under nitrogen.
3. Add 0·5 ml of methylene chloride and evaporate to dryness to ensure the azeotropic removal of water.
4. Dissolve the residue in 1 ml of benzene and make alkaline with 3—4 drops pyridine/benzene (1 : 4 by volume).
5. Make a derivative of the ethambutol and the internal standard by adding 20 μl trifluoracetic anhydride (sequanol grade) and standing for 2 hours at room temperature.
6. Remove excess of the derivatizing agent by washing into the aqueous phase using 0·01 M HCl.

7. Inject an aliquot $(2-3\,\mu l)$ of the benzene layer into a gas-liquid chromatograph equipped with an electron capture detector, which has a glass column, 6 ft $\times$ $\frac{1}{8}$ in, 3% OV-17 on Gas Chrom Q, 100–120 mesh; a carrier gas (nitrogen) flow rate of 20 ml/min; injector temperature 210 °C; oven temperature 157 °C and detector temperature 230 °C. The concentration of ethambutol can then be calculated by measuring the ratio of the peak heights of the derivatives of ethambutol and the internal standard (retention times approximately 4 and 2·5 minutes, respectively) and reference to a standard curve prepared using aliquots of plasma containing 0·1 and 1 μg ethambutol.

Other Assay Methods

The following assays are rarely used, and for this reason only the references for each method are given.

Determination of p-Aminosalicylic Acid

In serum and body fluids: Newhouse and Klyne (1949).
In urine: Case (1961).

Detection of Cycloserine and Pyrazinamide

In urine: Krauss et al. (1961).

Detection of Pyrazinamide

In plasma: Ellard (1969).

Detection of Ethionamide

In urine: Venkataraman et al. (1967).

Detection of Thiosemicarbazone

In urine: Short (1961).

References and Bibliography

Aronson J. D. (1926) *J. Infect. Dis.* **39**, 314.

Baker F. J. (1967) *Handbook of Bacteriological Technique.* London, Butterworths.

Bonicke R. and Lisboa B. P. (1959) *Zentralbl. Bakteriol. Reihe A* **175**, 403.

British Thoracic and Tuberculosis Association (1971) *Tubercle* **52**, 1.

Case C. M. (1961) *Tubercle* **42**, 531.

Castets M. (1969) *Médicine d'Afrique Noire* **4**, 321.

Collins C. H. (1962) *Tubercle* **43**, 292.

Corper H. J. and Stoner S. B. (1946) *J. Lab. Clin. Med.* **31**, 1364.

Dhayagude R. G. and Shah B. R. (1948) *Indian J. Res.* **36**, 79.

Dubos R. (1953) *The White Plague.* London, Gollancz.

Dubos R. and Davis B. D. (1946) *J. Exp. Med.* **83**, 409.

Dymond L. C. and Russell D. W. (1970) *Clin. Chim. Acta* **27**, 513.

Eidus L. and Hamilton E. J. (1964) *Am. Rev. Respir. Dis.* **89**, 587.

Ellard G. (1969) *Tubercle* **50**, 144–158.

Fernstein M. (1945) *Science* **101**, 675.

Friedman (1903) *Zentralbl. Bakteriol. Reihe A* **34**, 647–793.

Grange J. (1979) *Br. J. Hosp. Med.* 547.

Hauduroy P. (1955) *Derniers Aspects du Monde des Mycobacteries.* Paris, Masson et Cie.

Hansen G. H. A. (1874) *Nor. Mag. Laegevidensk.*

Health and Safety at Work Act 1974.

Jensen K. A. (1932) *Zentralbl. Bakteriol. Parasiten kd. Abt. 1, Orig.* **125**, 222.

Jensen K. A. (1955) *Bull. Int. Union Tuberc.* **25**, 89.

Karassova V., Weissfeiler J. and Krasznay E. (1965) *Acta Microbiol. Acad. Sci. Hung.* **12**, 275.

Koch R. (1882) Die Aetiologie der Tuberkulose. *Berl. Klin. Wochenschr.* **19**, 221.

Kraus P., Krausova E., and Simane Z. (1961) *Tubercle* **42**, 516, 521.

Kubica G. P. (1978) *Bull. Int. Union Tuberc.* **53**, 3, 192.

Kubica G. P. and Pool L. (1960) *Am. Rev. Resp. Dis.* **81**, 387–91.

Kubica G. P., Kaufman A. J. and Dye W. E. (1964) *Am. Rev. Respir. Dis.* **89**, 284.

Kubica G. P., Dye W. P., Cohn M. L. and Middlebrook G. (1963) *Am. Rev. Respir. Dis.* **87**, 775.

Lee C. S. and Benet L. Z. (1976) *J. Chromatogr.* **128**, 188.

Linell L. and Norden A. (1954) *Acta Tuberc. Pneumol. Scand.* **33**, 1.

Lloyd J. and Mitchison D. A. (1964) *J. Clin. Pathol.* **17**, 622.

Lorian V. (1966) *Am. Rev. Respir. Dis.* **94**, 459.

Lorian V. and Lacasse M. L. (1967) *Dis. Chest* **51**, 275.

Marks J. (1958) *Mon. Bull. Minist. Health* **17**, 194.

Marks J. (1963) *Mon. Bull. Minist. Health.* **22**, 150.

Marks J. (1964) *Tubercle* **45**, 47.

Marks J. (1965) *Tubercle* **46**, 65.

Marks J. (1965) *Br. Med. J.* **1**, 32.

Marks J. and Richards M. (1962) *Mon. Bull. Minist. Health* **21**, 200.

Marks J., Jenkins P. A. and Tsukamura M. (1972) *Tubercle* **53**, 210.

Ministry of Health (1958) *Monthly Bulletin* **17**, 99.

Mitchison D. A. (1966) *Tubercle* **47**, 289–91.

Mitchison D. A. and Spicer C. C. (1949) *J. Gen. Microbiol.* **3**, 184.

Mitchison D. A., Wallace J. G., Bhatia A. L. et al. (1960) *Tubercle,* **41**, 1.

Newhouse J. S. and Klyne W. (1949) *Biochem. J.* **44**, 7.

Ogawa T. and Sanami K. (1949) *Kekkaku* **24**, 2, 13.

Petroff S. A. (1915) *J. Exp. Med.* **21**, 38.

Prissick F. H. and Masson A. M. (1956) *Can. Med. Assoc. J.* **75**, 798.

Runyon E. H. (1955) *Am. Rev. Tuberc. Pulm. Dis.* **72**, 866.

Runyon E. H. (1959) *Bull. Int. Union Tuberc.* **29**, 69.

Schaefer W. B., Wolinsky E., Jenkins P. A. et al. (1973) *Am. Rev. Respir. Dis.* **108**, 1320.

Schwabacher H. (1959) *J. Hyg. Camb.* **57**, 57.

Selbie F. R. and O'Grady F. (1954) *Br. J. Exp. Pathol.* **35**, 556–65.

Short E. I. (1961) *Tubercle* **42**, 524.

Skerman V. B. D., McCowan V. and Sheath P. H. R. (1980) *Int. J. Systemic Bacteriol.* **30**, 225.

Stonebrink B. (1958) *Acta Tuberc. Scand.* **35**, 67.

Sula L. (1947) *Public Health Rep.* No. 63, 867–83.

Tsukamura M. (1975) Published by the Research Laboratory of the National Sanatorium, Chubu Chest Hospital, Obu, Aichi-ken, Japan.

Tsukamura M. and Tsukamura S. (1964) *Tubercle* **45**, 64.

Venkataraman P., Eidus L., Tripathy S. P. et al. (1967) *Tubercle* **48**, 291.

Villemin J. A. (1868) *Études Experimentales et Cliniques sur la Tuberculose.* Paris, Baillière et Fils.

Wayne L. G. (1961) *Am. J. Clin. Pathol.* **36**, 185.

Wayne L. G., Doubeh J. R. and Russell R. L. (1964) *Am. Rev. Respir. Dis.* **90**, 588.

W.H.O. (1963) *Bull. W.H.O.* **29**, 565.

Yates M. D., Collins C. H. and Grange J. M. (1982) *Tubercle* **62**, 55.

Index